Digital Exposure

DOI: 10.1057/9781137312402

Other Palgrave Pivot titles

Peter Taylor-Gooby: **The Double Crisis of the Welfare State and What We Can Do About It**

Jeffrey Meyers: **Remembering Iris Murdoch: Letter and Interviews**

Grace Ji-Sun Kim: **Colonialism, *Han*, and the Transformative Spirit**

Rodanthi Tzanelli: **Olympic Ceremonialism and the Performance of National Character: From London 2012 to Rio 2016**

Marvin L Astrada and Félix E. Martín: **Russia and Latin America: From Nation-State to Society of States**

Ramin Jahanbegloo: **Democracy in Iran**

Mark Chou: **Theorising Democide: Why and How Democracies Fail**

David Levine: **Pathology of the Capitalist Spirit: An Essay on Greed, Hope, and Loss**

G. Douglas Atkins: **Alexander Pope's Catholic Vision: "Slave to No Sect"**

Frank Furedi: **Moral Crusades in an Age of Mistrust: The Jimmy Savile Scandal**

Edward J. Carvalho: **Puerto Rico Is in the Heart: Emigration, Labor, and Politics in the Life and Work of Frank Espada**

Peter Taylor-Gooby: **The Double Crisis of the Welfare State and What We Can Do About It**

Clayton D. Drinko: **Theatrical Improvisation, Consciousness, and Cognition**

Robert T. Tally Jr.: **Utopia in the Age of Globalization: Space, Representation, and the World System**

Benno Torgler and Marco Piatti: **A Century of *American Economic Review*: Insights on Critical Factors in Journal Publishing**

Asha Sen: **Postcolonial Yearning: Reshaping Spiritual and Secular Discourses in Contemporary Literature**

Maria-Ionela Neagu: **Decoding Political Discourse: Conceptual Metaphors and Argumentation**

Ralf Emmers: **Resource Management and Contested Territories in East Asia**

Peter Conn: **Adoption: A Brief Social and Cultural History**

Niranjan Ramakrishnan: **Reading Gandhi in the Twenty-First Century**

Joel Gwynne: **Erotic Memoirs and Postfeminism: The Politics of Pleasure**

Ira Nadel: **Modernism's Second Act: A Cultural Narrative**

Andy Sumner and Richard Mallett: **The Future of Foreign Aid: Development Cooperation and the New Geography of Global Poverty**

Tariq Mukhimer: **Hamas Rule in Gaza: Human Rights under Constraint**

Khen Lampert: **Meritocratic Education and Social Worthlessness**

G. Douglas Atkins: **Swift's Satires on Modernism: Battlegrounds of Reading and Writing**

David Schultz: **American Politics in the Age of Ignorance: Why Lawmakers Choose Belief over Research**

G. Douglas Atkins: **T.S. Eliot Materialized: Literal Meaning and Embodied Truth**

Martin Barker: **Live To Your Local Cinema: The Remarkable Rise of Livecasting**

Michael Bennett: **Narrating the Past through Theatre: Four Crucial Texts**

Arthur Asa Berger: **Media, Myth, and Society**

Hamid Dabashi: Being a Muslim in the World

David Elliott: **Fukushima: Impacts and Implications**

Milton J. Esman: **The Emerging American Garrison State**

Kelly Forrest: **Moments, Attachment and Formations of Selfhood: Dancing with Now**

Steve Fuller: **Preparing for Life in Humanity 2.0**

Ioannis N. Grigoriadis: **Instilling Religion in Greek and Turkish Nationalism: A "Sacred Synthesis"**

Jonathan Hart: **Textual Imitation: Making and Seeing in Literature**

DOI: 10.1057/9781137312402

palgrave▸pivot

Digital Exposure: Postmodern Postcapitalism

Raphael Sassower
University of Colorado, Colorado Springs

palgrave
macmillan

DOI: 10.1057/9781137312402

First published 2013 by
PALGRAVE MACMILLAN

Palgrave Macmillan in the UK is an imprint of Macmillan Publishers Limited, registered in England, company number 785998, of Houndmills, Basingstoke, Hampshire RG21 6XS.

Palgrave Macmillan in the US is a division of St Martin's Press LLC, 175 Fifth Avenue, New York, NY 10010.

Palgrave Macmillan is the global academic imprint of the above companies and has companies and representatives throughout the world.

Palgrave® and Macmillan® are registered trademarks in the United States, the United Kingdom, Europe and other countries.

ISBN: 978–1–137–31241–9 EPUB
ISBN: 978–1–137–31240–2 PDF
ISBN: 978–1–137–31239–6 Hardback

This book is printed on paper suitable for recycling and made from fully managed and sustained forest sources. Logging, pulping and manufacturing processes are expected to conform to the environmental regulations of the country of origin.

A catalogue record for this book is available from the British Library.

A catalog record for this book is available from the Library of Congress.

www.palgrave.com/pivot

DOI: 10.1057/9781137312402

In memory of Jean-François Lyotard

whose version of postmodernism inspired me to practice it for decades

DOI: 10.1057/9781137312402

Contents

Preface vii

1 The Digital Evolution in Postmodernity 1
- I Beyond binary oppositions 2
- II Beyond any two cultures 5
- III Digital technophobes and technophiles 8

2 Postcapitalism: Materialism, Decentralization, and Globalization 19
- I A materialist approach 20
- II Decentralization/centralization 26
- III Globalization/poverty 31

3 Political Economy: Freedom, Surveillance, and Entrepreneurship 35
- I Freedom 36
- II Surveillance 39
- III Entrepreneurship 46

4 Legality and Morality: Intellectual Property, Virtual Currency, and Corporate Responsibility 51
- I Intellectual property: rights/duties 52
- II Virtual currency 56
- III Corporate responsibility 61

Conclusion 66

Selected Bibliography 72

Index 80

DOI: 10.1057/9781137312402

Preface

When different theories vie for legitimacy and relevance in the Digital Age, there is an ongoing suspicion that postmodernism falls short. The challenge has been most directly articulated by James Boyle who has lamented about the sad state of this way of thinking, especially in light of the "world of electronic media." More specifically, Boyle suggests that "one would expect postmodern philosophy to have sophisticated and interesting things to say about the ways in which a new political economy of information power is constructed, conceived, and defended" (Boyle 2008, Preface).

By and large, this book tries to take up this challenge unapologetically. The focus of my narratives remains political economy; the stage is the Digital Age; the playwright represents a hybrid approach that includes the materialism of Marxism, the critical rationalism of Popperianism, and the pluralism of postmodernity; and the actors speak for themselves through their texts. The play is supposed to be engaged and reflective, without any pretense to cover all there is to say about the *Digital Age*. The play ends up being performed in one form or another in a casino, as the Conclusion suggests, because there, more so than in many other market-like environments, one can test ideas about the relationship between individuals and the government as well as between equality and freedom. With uncertain expectations, the question remains: do all participants have equal opportunity to succeed?

In addition, this book's focus on political economy is informed by the Enlightenment Age of the eighteenth century (especially Adam Smith) when politics as the study of power relations and the structure of political institutions was understood to be intricately related to economic structures and principles about human behavior in the marketplace. The marketplace, just as any other arena of human interaction, assumed a level of equal access to information (about supply and demand, and therefore prices) so that individual actors had a sense of equal opportunity to participate. But, as Boyle reminds us in Karl Marx's terms, this wasn't the case (Ibid., ch. 3, 176–178). Perhaps what is at issue is under what conditions can equality still be claimed and in what domain, noting the new formation of information "classes" where marketplace entrepreneurship is curtailed by legal restrictions on behalf of corporate elites (Ibid.). Given the lobbying power of commercial entities, should standard economic models based exclusively on scarcity of resources, labor, and dreams be *displaced* by models based on digital abundance (data storage space, energy, imagination, creative proliferation)?

The third point of this book relates to the ways in which *postcapitalism* draws on what deserves retaining from both Smith and Marx: the only way to a just world is through the transformation of economic conditions. For Smith it was individual freedoms of choice with private property that leads to prosperity, while for Marx it was collective ownership of the means of production with individual protection and prosperity. They both acknowledged the material conditions under which change was possible for an ever-expanding capitalism or prosperous socialism. Industrial (nowadays digital) progress was neither the root of all evil nor the harbinger of all hope; instead, it could be harnessed to ensure prosperity for all. Both visions never fully materialized. Instead, we are at a point of postcapitalism where *hybrid* institutions struggle to balance individual choices (rights language) with social planning (welfare safety nets), as I have detailed elsewhere (Sassower 2009). The main point of postcapitalism is corrective so as to fine-tune classical ideals within a digital reality.

The fourth (and related) point focuses more specifically on the techno-scientific nexus that dominates postcapitalism. Apple does not make the economy, nor is Microsoft exclusively a byproduct of American capitalism. Instead, there is a critical mediation in the marketplace (in the form of tax policies or intellectual property rights), where certain economic conditions foster particular developments that in turn transform the marketplace: it's a hybrid political economy, drawing equally from

DOI: 10.1057/9781137312402

classical capitalism as from classical socialism, maintaining overlapping Enlightenment principles, such as individual liberty and equality. And, of course, multiple ethical issues are raised as well. Acknowledging the great differences between the European Union and the US, postcapitalism as a hybrid political economy can be examined in the ever-changing landscape of the Digital Age where no universal principles remain dominant.

What makes the political dimension of the Digital Age so powerful is the fact that in its materiality – gadgets and computers, Internet and websites – it challenges default choices of citizens: should they ignore a political scare? Can they retreat to a cocoon and tune out the digital cacophony surrounding them? At some point in the evolution of digital technologies we find gradual acceptance and rejection of these technologies: we allow more of this and less of that, we begin to focus on filters to direct our attention to what seem to us significant moments, and when doing so, we recognize how misleading the filters may be as well as our reliance on gate-keepers to navigate our choices. Some claim that we are chained to our devices and work much harder than ever before, even on scheduled vacation (Lanier 2010), becoming more alienated from each other than ever before (Turkle 2011); others suggest that the mobility of cell phones (personal computers, iPads, and the like) allow us to navigate work/leisure time under much more favorable conditions than ever before (Chorost 2011, Powers 2010); still others are unclear what boundaries ought to be set between the private and public in a digitally transformed age (Morozov 2011). Digital technologies foist on us in the starkness of their materiality an inescapable set of choices that are political in the sense that they are drawn against a particular ideological backdrop (democratic, populist, egalitarian, or paternalistic), and as such are more invasive than print media, for example, or political stump-speeches.

Fifth, this book examines the technophile–technophobe divide as it relates to the Digital Age. The promise of a change in our personal and collective mind-set because of the digital tools that have advanced since the 1970s with the advent of computers and more thoroughly with the access to the Internet in the 1990s can be understood as a historical "game-changer." According to Marxism, the material conditions of an era, may it be feudalism or capitalism, determine the society's superstructure: the kind of political and social institutions that reflect those material conditions. Moral and legal sanctions, including but

DOI: 10.1057/9781137312402

not limited to Internet surveillance by private and government entities, accompany radical digital transformations. Contemporary writers explain, for example, that the Digital Age has brought about access to music in a manner that has changed young people's view of intellectual property and copyright. Given easy downloads, the idea of having to pay for music seems odd just as much as the idea that it's wrong to liberally quote and re-mix without any attributions (Lessig 2008). And the idea of "the author" is romantic at best, and at worst provides legal cover (intellectual property) to stifle innovation (Boyle 2008). The point is simple: digital conditions force us to rethink ideas, knowledge, information, and the democratic institutions that protect their production, distribution, and consumption.

But is it indeed this simple? What can we say about the transformation of human personality (see Aboujaoude 2011, on e-personality)? It's a question of dialectical and reciprocal influences, where one set of conditions informs another and is likewise informed by them. This formulation allows for a shift from a *hypothetic-deductive thinking* derived from physics (Rosenberg 1988), which remains *reductionist* at its very core, to more contemporary thinking about *emergent qualities* derived from biology (Mayr 1997) that are more open-ended. Emergent qualities allow for more nuanced, open-ended conclusions, because they are not causally predictable, and in their unpredictability, they provide for a reassessment of the set of conditions (interior and exterior, genetic and environmental) that bring them about. In this context, Smith's concern with the *Impartial Spectator* comes to mind: a social lubricant brings forth a sense of social cohesion and communal concern for the well-being of all its constituent members (Smith 1976/1759). Whether understood in terms of moral sentiments inherent in humanity or in religious terms, Smith could seventeen years later more easily introduce the *Invisible Hand* that guides marketplace exchanges (Smith 1937/1776). This way of presenting economic institutions is less reductionist (finding a set of causes that necessarily leads to particular consequences) or universal (relying on some notion of human nature and the fixed ways in which humans interact). Instead, layering one set of economic and legal conditions on another layer of social and moral conditions illustrates their mutual reliance and reciprocal influence.

My *first caveat* is that the version of postmodernity I endorse (primarily associated with Jean-François Lyotard) promotes the level playing field for all individual discourses, empowering them in the name of plurality

DOI: 10.1057/9781137312402

and multiplicity. When faced with digital technologies, one may ask: isn't this process itself threatened to be homogenized and organized by diffuse power-sites whose identities remain secret and whose agendas unknown? Has capital formation, concentrated in the hands of virtual individuals (in their virtual financial sites), displaced power formation from the hands of monarchs and dictators lodged in castles? Celebrity owners of large shares of Microsoft, IBM, Apple, Oracle, Facebook, and the like wield more power than some heads of state; their control is not limited to the development and production of technologies, but extends to their distribution and consumption and therefore to politics. This, then, isn't the standard view of postmodernism as a relativistic, "anything goes," irresponsible stand.

The *second caveat* is that the philosophical orientation here is at once alarmed by the false promises of modernism, postmodernism, and technoscience because they tend to ignore political–economic structures, and in their respective ways hyped the digital revolution as a game-changer. Yet, I nonetheless use all three to figure out pragmatic solutions to fundamental problems. If the democratic state, as a custodian of individual rights and dreams, is relegated to the role of accommodating digital giants, allowing unbridled accumulation of wealth and power, then there is little hope that the ideals of both Smith and Marx will ever come to fruition. It's not the state versus the individual, collectivism versus individual property; instead, it's the state working on behalf of and in the interest of individuals to ensure a fair marketplace for goods and ideas. The Soviet failure has taught the European Union what not to do, and the European Union experiment gives pause to the Americas because of its failure to dictate economic policies in member states without a central banking authority. Perhaps a focus on digital technologies will clarify advantages and shortfalls these political economies need to address when experimenting with postcapitalist models.

The *third caveat* is that no claims for comprehensiveness are made here. Since the focus is more on political economy than anything else, important contributing factors may be underdeveloped, such as education, medicine, military and warfare, psychology, entertainment, and gaming. Given the prescribed scope of this book, they fall outside its range. Each one of them deserves its own play with its own stage and playwright; each of them is interwoven into my own play, but if fully developed would make the book unbearably long. By way of example, let me mention a recent publication by Eric Topol (2012) in which he appropriates Joseph

DOI: 10.1057/9781137312402

Schumpeter's claim that "creative destruction" denotes the "transformation that accompanies radical innovation" as we are seeing in medicine. Fascinating as the case of medicine may be, and as close as it may be coming to test the scope of the "digital revolution," this particular case fails to address broader economic and political issues. By focusing on patient care, assisted by computerized gadgets that provide an ongoing flow of information, Topol ignores fundamental critical disagreements about ideological frameworks that may derail if not obstruct such a utopian vision of the future of medicine. This is not to say that one cannot apply the insight of the present book to the case of medicine. On the contrary, once fully understood, this and many other cases and areas of application can be revisited. Perhaps they warrant another play; perhaps just another act; perhaps a revolution isn't as imminent. Time will tell. In the meantime, let's view this play without the distraction of so many other digitized plays.

* * *

I'd like to thank all those kind enough to read parts or the entirety of this manuscript. Professor Joseph Agassi, a teacher and wise counsel for years, made valuable comments and corrections. So did Professor Sam Gill who is more intimately conversant in this area than anyone I know. Professor Steve Fuller has been a guide and critic for years; his unfailing support means so much to me. Dr. Cindy Beggs, Moti Seuss, Eyal Kaplan, Jaime Reixach, and Garrett Coon read the whole manuscript and provided helpful feedback to ensure that the main argument would not be lost. And finally, Phillipa Grand, my editor, showed courage in allowing me to join her illustrious author lineup; I'm humbled by her trust in me.

When teaching a graduate seminar on Digital Culture at Tel Aviv University (2012) to students the majority of whom are part of the Israeli "high-tech" industry, I realized that their interest in postmodernism as a way to think about the cultural impact of digital technologies should be pursued.

And finally, I'm grateful to the University of Colorado at Colorado Springs for giving me the intellectual freedom to pursue my interests beyond my job description (if it still has it in its paper files). Only celebrities at well-funded think tanks enjoy the luxury we, tenured full-professors, take for granted.

1
The Digital Evolution in Postmodernity

Abstract: *Instead of the binary division between capitalism and socialism or between modernism and postmodernism, there is a way to distill the best (most efficient and socially beneficial) that these binaries can offer so as to find a pragmatic middle-ground. Digital technologies don't align more strongly with one side of the binaries, but offer a way to realize and test some elements of both of them. In doing so, an evolutionary trajectory is outlined so as to acknowledge the continuity rather than revolutionary nature of the Digital Age.*

Sassower, Raphael. *Digital Exposure: Postmodern Postcapitalism*. Basingstoke: Palgrave Macmillan, 2013. DOI: 10.1057/9781137312402.

I Beyond binary oppositions

In order to examine the potential of a multilayered approach to the Digital Age in the postmodern world whose postcapitalist marketplace encourages participation and prosperity, we should break down some conceptual barriers that stand in the way of clear thinking. Among them is a complacent confinement to traditional binaries, such as modernism and postmodernism, capitalism and socialism, and technoscience and the humanities. Each binary has its justification for drawing distinctions so as to classify some ways of thinking as modern as opposed to postmodern, or ways of conducting business as capitalists rather than as socialists. Though radically different, such binaries entrench differences among ways of thinking rather than seeing all proposals as fluid and open-ended and amenable to change when changing circumstances demand a change in approach.

The postmodern condition, so eloquently outlined by Jean-François Lyotard (1984), has been with us all along and therefore any sense of historical sequencing (modern→postmodern) misses the point of the postmodern condition. Lyotard explains the power relationship of modernity and capitalism colliding for universal hegemony. One way to illustrate the parallel, yet different, modes of thinking that distinguish modernism from postmodernism can be seen in Table 1.1 below. Note that at times the two columns complement each other (both and) while at others they are oppositional if not outright contradictory (either or).

Table 1.1, which is two-columned, attempts to capture the differences that inform the two ways of thinking in the most general manner and without any pretense of comprehensiveness. Yet, it illustrates the kind of conceptual reference-points worthy of examination (see an alternative in Hassan 1987). To some extent, we all live in both "camps" or inadvertently participate in or make use of these two ways of thinking, depending what we are dealing with. In the Digital Age, a recognition of these intertwined mind-sets may assist us in figuring out why we seem confused when computer engineers are closed-minded in one sense (within their code) and open-minded in another (sharing their work with others free of charge). The postmodern condition finds allies in every technological age, from experiments during the Industrial Revolution to contemporary attempts at reconfiguring intellectual property rights (see Berry 2008, Boyle 2008). If we layer these two

DOI: 10.1057/9781137312402

TABLE 1.1 *The Postmodern Condition*

MODERNISM	POSTMODERNISM
Certainty (science)	Ambiguity (the human condition)
Absolutism (values, morals)	Relative Contexts/Framing
Rationality/Logic	Reasonableness (modified choices)
Foundation (physical, legal)	Context (mediated differences)
Determinism (history)	Indeterminism (chance, unpredictability)
Legitimacy (God, nature)	Self-legitimation (Declaration of Independence)
Hierarchy (Church, State)	On-par Plurality (minority voices)
Meta/Grand-Narrative (history)	narratives (histories)
Singularity (purpose, opinion)	Multiplicity (differences)
Homogeneity (universality)	Heterogeneity (differences)
Replacement (sequentially)	Displacements (concurrent developments)
Rules/Laws (absolute)	Case-by-Case (circumstances)
Truth (one and only)	Putative truths (tentative until proven false)
Meaning (transparent)	Discourses (opaque, incommensurable)
Spectator (outside observer)	Participants (part of nature's measurements)
Freedom (absolute, universal)	Liberties (constrained, circumscribed)
Equality (absolute, universal)	Equal Opportunity (conditional, circumstances)

TABLE 1.2 *Capitalism and Socialism*

CAPITALISM	SOCIALISM
Economics	Political Economy
Laissez-Faire	Planning (five-year plans)
Competition	Cooperation
Money (only measure)	Honor/Dignity
Individual Self-Interest	Collective Consciousness
Division of Labor	Prescribed Duties
Supply and Demand (law)	Central Planning
Market Efficiency	Inefficiencies
Transparency/Trust	Secrecy/Manipulation
Risk (calculated)	State Protection
Freedom/Equality	Power Relations
Private Property	State Ownership of Property

DOI: 10.1057/9781137312402

mind-sets and practices with those of political economy (capitalism and socialism, as in Table 1.2), we can more fully appreciate the complexity on one hand and the complete unpredictability on the other of this new emerging age of digital technologies.

With the outlines of what capitalism and socialism ought to look like in their ideal states, we should continue to speculate about their potential for sustainable durability and success. Capitalism has traditionally been couched within a legal system that protected certain rights and activities through tax legislation, such as deduction for investments and losses, or through the courts to enforce contracts. As seen most recently in the US, "too big to fail" has been a mantra that justified the use of enormous public funds to help prop commercial and investment banks alike whose bad decisions would have brought them to bankruptcy. The European Union is likewise saddled with national capitalist markets and social programs that require intervention and subsidies unknown to the drafters of ideal capitalist markets. The same judgment is appropriate in the case of socialism, since the best we have seen in the past one hundred years are cruel examples of state communism (USSR, China, and many others) that deliberately overlooked the humanistic foundations of ideal socialism where overall abundance (post-capitalist wealth accumulation) would indeed cater to everyone's needs and everyone would eagerly contribute to the general welfare of all.

Absent these idealized states, a compromised political economy has presented itself. Under these practical circumstances we are witnessing mixed-economies where welfare commitments are available but resented, or where a central authority dictates business practices. The notion of postcapitalism, as we shall see in more detail later, therefore is neither alien nor worrisome, but rather a more formalized restatement of trends that we should endorse, promoting the best both capitalism and socialism offer in their pristine versions. It seems that the confusion of contemporary digital culture when fishing for ideas that have in fact been there all along is that in doing so it is unprincipled. But that is exactly what is at stake: *postmodern postcapitalism, a hybrid mode of being that celebrates the best it finds without regard to traditional boundary conditions or outdated restrictions*. Yet, amidst this celebration it's only fitting to acknowledge a troubled history that we have to overcome in the Digital Age, the dividing lines between technoscientists and humanists.

DOI: 10.1057/9781137312402

II Beyond any two cultures

Though the horrors of the twentieth century are behind us – world wars, atomic bombs, genocides – their residues still affect our thinking. One of them is the troubling distance between the humanities and the sciences, what C. P. Snow famously called the "two cultures." Can digital technologies ameliorate this concern? And if they do, when artists and computer scientists collaborate on video games, does this collaboration indeed bridge their training and conceptual divide? According to I. I. Rabi (1970), one way to bridge the divide is not simply to find common intellectual grounds or practical reasons, such as the politics of the Cold War of yesteryear or video games today, but through an affirmation of human wisdom. Human wisdom commands us to make use of whatever intellectual knowledge we can gather from every available resource and test it against our experiences to produce effective and life-saving policies. Common educational structures should be in place to accommodate this process, whether understood in the pragmatic tradition of the American John Dewey or in the more abstract sense of the March of Reason as propounded in the nineteenth century by the German philosopher Hegel. The goal, however achieved, is to bridge the intellectual and socioeconomic gaps that have become evident by the middle of the past century (in an age of expertise, Sassower 1993) to ensure the survival and prosperity of the human race.

Some of the concerns related to the two cultures remain intact: the technical prowess of scientists and engineers in the Digital Age seems magical even in the eyes of literary intellectuals who use digital gadgets. But while the "cultures" discussed in the past century were privileged insofar as they embodied the intellectual and aristocratic classes of the UK, the cultures we observe in twenty-first century America are much more bifurcated. We have the educated and the less educated cultures (in terms of postsecondary education), and within each one of these we have distinctions associated with literacy and expertise (cutting-edge digital technologies) and areas of work and thought (from entertainment to politics). Would Snow or any of his contemporaries associate television-literate audiences with what he perceived to be "literary intellectuals"? For him, intellectuals were academically educated and at times employed within the academy; some were interested in literature and some in the sciences. He was struck by the fact that they actually had much more in common than they themselves would admit, and it's exactly that fact he

DOI: 10.1057/9781137312402

wanted to exploit to bridge the widening gap between the sciences and the humanities.

What seemed a reasonable expectation in the past century may seem ludicrous in this one, when the gap between the rich and the poor is ever-increasing, when higher education funding is undermined in part because of the poor economic conditions of many states that traditionally subsidized their namesake universities (for example, the state of Colorado provides less than 5 percent of the annual budget of the University of Colorado system), and when the overall anti-intellectualism of the nation, as Richard Hofstadter illustrated (1966), keeps rearing its ugly head. It's not only that the so-called two cultures are in fact ten or twenty different cultures (science/humanities, rich/poor, educated/uneducated, rural/urban, religious/secular, visually literate/illiterate, married/single, healthy/sick, young/old, politically engaged/disengaged, and so forth), but that they are socioeconomically defined, without a shred of a national agenda. Do we find common ground when dreaming of world peace? Is our singular national goal greater prosperity? Is that prosperity circumscribed by digital innovations? And will it be spread fairly across all socioeconomic classes?

One proposal for bridging the so-called divide between the sciences and the humanities, according to Friedrich Kittler, claims that there really isn't a divide at all (so no bridge is necessary): everything worth studying must be measured, counted, and be given proportions. All areas are reducible into one nexus. In his words: "the only things that can be known about the soul or the human are the technical gadgets with which they have been historically measured at any given time" (Kittler 2010, 35). In light of this view that bridges the humanities and the sciences, there is a historical development, following Vilem Flusser, that allows Kittler to show how we have moved from the four-dimensionality of space and time to the three dimensions of obelisks, then to two dimensions of gravestones, to the one-dimensionality of text or print, all the way to the zero-dimensionality of numbers and bits in computers so as to avoid "any danger of concealment whatsoever" (Ibid., 226–227). What this process reveals is a deep commitment to reductionism – the view that multiple effects or events can be retrospectively reduced to a single antecedent or cause – as a way to make sense of the linear progression of technoscientific developments and the cognitive reflection that attends them. Unlike Snow, Kittler identifies only one culture.

DOI: 10.1057/9781137312402

In Kittler's words: "When seen from this perspective, computers represent the successful reduction of all dimensions to zero. This is also the reason why their input and output consisted of stark columns of numbers for the first ten years after 1943. Operating systems like UNIX introduced the first one-dimensional command lines in the sixties, which were then replaced by a graphic or two-dimensional user interface in the seventies, beginning with the Apple Macintosh. The reason for this dimensional growth was not the search for visual realism, but rather its purpose was to open up the total programmability of Turing machines at least partially to users, which demands as many dimensions as possible due to the inconceivable number of programming possibilities" (Ibid., 227). For him, the transformation into fewer dimensions that then reverses into more dimensions is linked to human agency with specific needs for representation and control as means for comprehension and empowerment. He continues: "The transition to three-dimensional user interfaces (or even four-dimensional ones if time is included as a parameter), which today goes by the phrase 'virtual reality,' can of course also be understood as an expansion of the operational possibilities. Virtual realities allow for the literal immersion of at least two distant senses, the eye and the ear, and at some point they will also enable the immersion of all five senses. Historically, however, they did not originate from the immanence of the development of the computer, but rather from film and television" (Ibid.)

As illuminating as this rendition of the underlying similarity between the humanities and the sciences, there is also another layer of exposition that guides this historical narrative and that is warfare (see a fascinating parallel discussion of the development of computers – Turing machines – in the US in the hands of those developing the atomic and hydrogen bombs during World War II, Dyson 2012). Kittler is committed, ideologically and intellectually, to reducing and classifying all historical stages in warfare terms so that warfare as a motive and goal overshadows any sense of curiosity or play (that lead to technoscientific discoveries). For him, this "bombardment of the senses" is associated with warfare and more specifically with bomber pilot trainees (Kittler 2010, 228). Whether or not warfare is the underlying guide to bridging the purported cultural division Snow and Rabi worried about in the previous century, in this genealogy Kittler sets the tone for the dread and euphoria associated with contemporary digital technologies.

DOI: 10.1057/9781137312402

III Digital technophobes and technophiles

Despite the great diversity in contemporary culture, a diversity that probably could be detected in previous generations if the investigatory net were cast widely and deeply enough, there might be something that could inadvertently bring almost all of us together: digital technologies. Astrophysicists and grounds-keepers alike enjoy the use of cell phones and Global Positioning Systems as well as other gadgets provided by banks and fast-food franchisers even though their understanding of the internal workings of these instruments might radically differ. The Digital Age in this respect binds our culture in ways that would have been inconceivable a century ago. This is not to say that the differences among the different sub-cultures have been therefore erased or bridged; only to say that perhaps digital technologies could provide the means by which such erasure or bridging may come about (without resorting to Kittler's reductionism).

According to David Berry, contemporary culture is a "*computational* knowledge society," (2011, 3, emphasis in original), one whose uniqueness is actualized in its materiality (real storage units that emit heat and need to be cooled off) rather than some sense of its amorphous immateriality of the so-called cloud where data are stored or where computational activities take place. Following Martin Heidegger and Steve Fuller, he insists on the "concrete thing-in-the world-ness" of our computational culture (Ibid., 10). Overcoming the dichotomy between modernity and postmodernity, Berry suggests a difference between instrumental rationality and computation, so that "computational rationality is a form of reasoning that takes place through other non-human objects, but these objects are themselves able to exert agential features, either by making calculations and decisions themselves, or by providing communicative support for the user" (Ibid., 13). He calls this process without an end "an agonistic form of communicative action where devices are in a constant stream of data flow and decision-making which may only occasionally feedback to the human user" (Ibid., 14). Berry also emphasizes the importance of "digital intellect" rather than "digital intelligence" so as to add the critical dimension of the digital use with an eye on evaluating meanings rather than simply data in the context of "digital literacy," hence moving the discussion way beyond the binary oppositions listed above to a new terrain that characterizes the twenty-first century (Ibid., 20).

DOI: 10.1057/9781137312402

What the written word and eventually the printing press were supposed to accomplish for modernity, digital technologies might accomplish for postmodernity: tools for enlightenment and greater happiness. This is the story of technophiles, the lovers of all things technical, who believe in the emancipatory power of technoscience. Perhaps they are simply echoing the belief in reason, a reasonable belief, since its claim for universality bridges all gaps of tradition, ethnicity, religion, and socioeconomic stratification. If an argument is valid, Aristotle already offered centuries ago, then no personal bias comes into play. If technoscience teaches us that a fact is true – gravity or the relation between energy, mass, and the speed of light – then it's true for all of us, whether we are educated at Oxford or herd sheep in the hills of Afghanistan. This amazing revelation has carried the torch of enlightenment since the great scientific revolutions of the sixteenth and seventeenth centuries. It's because of this single attribute of technoscience that the great humanists, as Snow called them, had to find a way to address the entire human enterprise, rather than worrying, as the British have for centuries, about their particular class distinction, or the Germans about their own unification after the fall of the Berlin Wall in 1989. But just as digitized technoscience has its promise, so do the humanists share a concern for life and its meaning, as articulated by the existentialists for two centuries, as well as with human relations, such as friendship and love.

Members of the Frankfurt School, from Walter Benjamin and Theodor Adorno to Max Horkheimer and Herbert Marcuse, all the way to Jürgen Habermas, agree that a neo-Marxist critique of the advancement of technoscience in its capitalist trappings is essential in order to rescue civilization from its own abyss. It's easy to confuse the critique with a strong sense of technophobia – a deep distrust of technoscience and digital technologies, almost a fear that they will push humanity beyond any potential redemption. But only if this critique is read as a Messianic protest will it be understood negatively. From Benjamin's concern with the mechanical reproduction of works of art (1936) to Marcuse's concern with the one-dimensionality of humanity (1964), it becomes clear that the critique is in fact a *manifesto for salvation*: humanity can find its soul again. This, perhaps, is the spirit of Heidegger who worried that the advent of technology would doom us to the machinery of our surroundings instead of the pristine comfort and authenticity of the Black Forest of his youth (1954). A contemporary critique, such as Kittler's, is less romantic and more devastating, especially as he intertwines the onward march of mechanization and computation with warfare. For example,

DOI: 10.1057/9781137312402

"Once storage media can accommodate optical and acoustic data, human memory capacity is bound to dwindle. Its 'liberation' is its end" (Kittler 1999, 10). Or, put in another way: "Mechanization relieves people of their memories and permits a linguistic hodgepodge hitherto stifled by the monopoly of writing... The epoch of nonsense, our epoch, can begin" (Ibid., 86). Or, even more critically: "The age of media (not just since Turing's game of imitation) renders indistinguishable what is human and what is machine, who is mad and who is faking it" (Ibid., 146).

Unlike the Frankfurt School, Kittler's reactionary modernism is almost postmodern in its liberal juxtapositions: "The history of the movie camera thus coincides with the history of automatic weapons. The transport of pictures only repeats the transport of bullets... Nothing, therefore, prevented the weapons-system movie camera from aiming at humans as well. On the three fronts of war, disease, and criminality – the major lines of combat of every invasion by media – serial photography entered into everyday life in order to bring about a new body" (Ibid., 124–128). "The typewriter became a discursive machine-gun. A technology whose basic action not coincidentally consists of strikes and triggers proceeds in automated and discrete steps, as does ammunitions transport in a revolver and a machine-gun, or celluloid transport in a film projector" (Ibid., 191). He summarizes his historical tour-de-force: "In order to supersede world history (made from classified intelligence reports and literary processing protocols), the media system proceeded in three phases, Phase 1, beginning with the American Civil War, developed storage technologies for acoustics, optics, and script: film, gramophone, and the man-machine system, typewriter. Phase 2, beginning with the First World War, developed for each storage content appropriate electric transmission technologies: radio, television, and their more secret counterparts. Phase 3, since the Second World War, has transferred the schematic of a typewriter to a technology of predictability per se; Turing's mathematical definition of computability in 1936 gave future computers their name" (Ibid., 243). Just like the critique of the Frankfurt School, this so-called postmodern critique benefits from being *immanent* insofar as it finds problematic as well as redemptive qualities in the very presence of the apparatus depicted as the shackles of modernity. Instead of being *this or that*, technoscientific modernity *or* authentic humanity, it becomes *this and that*, a postmodern blend that encompasses the worst and best humanity has produced.

When speaking of the postmodern environment in which yet another technoscientific wave is engulfing us, it should be made clear what this

DOI: 10.1057/9781137312402

means. For some, postmodernity has nothing to offer except political and moral vacuum that justifies whatever one does. It is therefore a retreat from the serious foundation enshrined by modernity, with Kantian principles and categorical imperatives (1981), and other Enlightenment ideals that find their ways into legal systems, such as the US Constitution. For others, it is a way of thinking, a mind-set, or a condition that opens up new horizons and empowers the disenfranchised to voice their views and grievances so as to transform hierarchical structures into collegial and mutually respectful ones, overcoming all the disadvantages of the past (Lyotard 1984). For still others, the very openness of pluralism and multiplicities decentralizes a whole set of political and moral concerns so that despite their promise of greater equality and liberty they may turn out to be practically disappointing and powerless (Baudrillard 1983).

Within this postmodern reality the new wave of digital technologies finds a different audience. It's not only that the technologies are radically different from what, for example, washing machines and television sets brought to Western cultures, but more importantly, the settings have radically changed as well. It's one thing to introduce washing machines to heterosexual households where there is a "male bread-winner" and a wife who takes care of household chores, so that this introduction reconfigures time constraints on household work and wives are now "free" to work in the traditional workplace (Chang 2010, "Thing 4"). Similarly, introducing television sets into middle-class households captures leisure time and other socializing activities (dinner, family communication, homework, and entertainment), regardless of their addictive powers (Mander 1978). It's quite another thing to introduce cell phones and the Internet to a heterogeneous population that counts among its households a majority of single-living as well as non-married cohabitation, single-mother heads of households, and the elderly living alone (Klineberg 2012). With this in mind, the impact of a new wave of digital tools and gadgets is exponentially less predictable.

If washing machines and electric irons, dishwashers and microwave-ovens, vacuum cleaners and ice-makers (as opposed to gramophones, films, and typewriters which Kittler enumerated) have all contributed to liberating household work from the burdens of past generations, they have done so in a predictable fashion (just as Kittler's favorites have advanced warfare). In traditional homes women worked less on the very chores that chained them to the home and so were free to pursue education and work, contributing to their own emancipation and the

DOI: 10.1057/9781137312402

welfare of their households. But can the same be said of the Internet and smartphones? The Frankfurt School's critique looms large over this question. There are some, like Jacques Ellul, who share some of their same concerns, but fashioned another set of sociological descriptions and solutions (1964/1954) because he sought a spiritual way for humanity to find "technical solutions" to overcome the ubiquity of the technological age. Then there are some, like Langdon Winner, who were concerned from a political perspective that technology is "out of control," and therefore ought to be structurally better organized to ensure the survival of the species (1977). There were also those, like Hans Jonas, who remind us of the ethical dimension that shouldn't be left out of the human–technological equation if we are to live a harmonious life, a life worth living (1984). The emphasis here is on personal and social responsibility, the kind that is loudly echoed by ecologists and environmentalists, fully summarized by Frederic Bender (2003).

Marshall McLuhan tried to capture some of what many critics tried to say in more sophisticated (and therefore less accessible) language when he asserted that "the medium is the message" (1994/1964, 7). He describes the "Age of Anxiety" as a time that "compels commitment and participation" because of the "electric implosion" that has brought about "the final phase of the extension of man." For him, "the technological stimulation of consciousness, when the creative process of knowing will be collectively and corporately extended to the whole of human society" leads to the realization that the medium is the message (Ibid., 3–5). McLuhan's insight is that "the personal and social consequences of any medium – that is, of any extension of ourselves – result from the new scale that is introduced into our affairs by each extension of ourselves, or by any new technology" (Ibid., 7). This phrasing may sound circular, but its significance is to point out that scale does matter: the scale of new digital technologies determines their "results" or "consequences." What is so frightening about digital technologies is not that they are the latest wave of ongoing technoscientific developments, but that their scale may overshadow centuries of incremental (and therefore comprehensible) changes. He is correct, of course, in diverting our attention from content concerns over technology (the actual objects and their messages) to its form, scale, and ubiquity.

What is unique about "digital culture," as Milad Doueihi eloquently suggests, is that it is "the *only* rival to religion as a universal presence" (2011, 3, emphasis in original). Recalling Ellul's spiritual concerns, he

DOI: 10.1057/9781137312402

struggles to explain what distinguishes this age from previous ones in addition to its obvious ubiquity: "The digital environment that envelopes us promises to be a thrilling adventure for the one who can read and remember as well as interact with the machine and its evolving interfaces, but an adventure that, despite all of its newness, recalls and extends well-known models of intelligibility and sociability" (Ibid., 11). Right away the shadows of Ellul (machines) and McLuhan (extension) loom large over this characterization, just as the ghosts of the Frankfurt School remain in the backdrop (unexpected newness). But what is more interesting is the fact that for Doueihi there is a "digital divide" that unlike Snow isn't between literature and science but based on socioeconomic differences: "Between those who have access to the various components of the digital environment and those who have not; between cultures and societies that are riding the wave of the new technologies and those that are either resisting it or left altogether outside its scope of influence for either political or financial reasons" (Ibid., 13).

For Snow the divide was almost purely intellectual, a choice some made to become literary scholars rather than scientists, and one made, as mentioned above, by an already privileged class of British academics and aristocrats (especially in light of earlier ages in which the two were the same). For Doueihi, by contrast, the divide isn't by choice, unless there is an ideological resistance; instead, the divide is an outcome of an existing bifurcation of socioeconomic conditions that allow some to fully enjoy the fruits of the digital environment, while others simply can't afford access or the literacy required of full membership in such an advanced and rich environment; this then sets in motion a hierarchy in digital literacy (Ibid., 35), reminiscent of Boyle's notion of information classes (2008). Since this is still a process of digitization – where the scope of digital technologies and their reach continues to expand – there are moments that appear as if we are still mixing old print media with digital ones, as a hybrid (Ibid., 19). Yet, such moments never fully overcome the onslaught of the digital culture as it overtakes old forms and modernizes them to ensure full participation and compliance, in short, control. There is no conspiracy in this process in the sense of a secretive, intentional, and systematic overtaking; instead, this is a natural process, an expected and unbridled evolution that cannot but continue in its current form of full transformation, ensuring a seamless shift from one developmental stage to the next.

The closest process, however deplorable in all its formation, that predates this one is the one described so frighteningly by Zygmunt Bauman

DOI: 10.1057/9781137312402

(1989). According to him, the same rational process that motivates the gardener to weed, distinguishing and separating weeds from flowers, pulling out weeds to ensure the health of the flowers, is the process that the Nazis followed when distinguishing between and separating the Aryans and Jews. In his words: "At no point of its long and torturous execution did the Holocaust come in conflict with the principles of rationality. The 'Final Solution' did not clash at any stage with the rational pursuit of efficient, optimal goal-implementation. On the contrary, *it arose out of a genuinely rational concern, and it was generated by bureaucracy true to its form and purpose*" (Ibid., 17, emphasis in original). Weeding out Jews to ensure the future of an Aryan supremacy is logical in this modern sense of classification according to a set of criteria that has a foundation (however demented) and leads to a hierarchy (however cruel and inhuman). It's the *logic of modernity* that gave intellectual legitimacy to this kind of thinking, according to Bauman, the kind of thinking that requires the full-fledged power of critique and dissent. Of course, not all instrumental rationality leads to genocide, nor is logic itself at fault for such acts of barbarity. But the point is that when the rational evolution of modernity expands its reach, it undermines the possibility of critique itself. And this lament, whether neo-Marxist, feminist, or postmodernist, comes forth to protest and provide alternatives.

Among the critics of digital technologies and the culture they breed is Jaron Lanier who argues that the collectivist philosophy of the Digital Age, what he calls "cybernetic totalitarianism," contains an imminent danger, which is bent on replacing individual thoughts with digital compliance, reducing people to mere units. He is beseeching us to think critically about the restrictions of technology to resolve problems of interpersonal deliberation and interaction. He begs us all to question the "personal reductionism" of our technological world, not simply in terms of addictive behavior but also in the ways we have been socialized to interact with our gadgets. Virtual reality, he contends, is attacking personhood (Lanier 2010). And that sentiment is echoed by many others who, while finding many positive effects of the digital culture, maintain that certain personality transformations are taking place at the neurological level (Brockman 2011).

Along these lines, there are some, like Aboujaoude, who claim that the Internet is unleashing our worst instincts by connecting us with gambling and prostitution websites, providing immediate gratification and uncontrollable impulse purchasing. By shielding the individual from

DOI: 10.1057/9781137312402

any exposure and encouraging a fabrication of who one is, the Internet licenses us to customize our fantasies and immediately act on them. As a psychiatrist concerned with interpersonal communication and its attendant pathologies, Aboujaoude finds everything about the Internet to be bad, from online compulsions to addictions. The “e-personality” is the kind of Freudian Id of infantile projections and impulses that needs to be curbed by the Superego. By probing into how one’s virtual reality affects one’s tangible being, he views personal traits as being mechanically launched into our offline existence. Drawing from his clinical work and personal experience, Aboujaoude illustrates how the disproportionate use of the Internet, smartphones, and other technologies can cause a direct change in our personalities, so that we become intolerant, impetuous, neglectful and egotistical (2011).

It’s this particular digital technology that has the power to transform our personality, according to Nicholas Carr. Technology amplifies everything, both good and bad instincts. Agreeing that technology itself is amoral and therefore in itself neither good or bad, what it does to our brains by rewiring them may short-circuit emotions, such as empathy. This occurs because of the speed of communication and the kind of short attention-span our brains become used to so that emotive filters may be ignored and nasty commentary easily dispensed. In his words: “As a technology, a book focuses our attention, isolates us from the myriad distractions that fill our everyday lives. A networked computer does precisely the opposite. It is designed to scatter our attention ... Knowing that the depth of our thought is tied directly to the intensity of our attentiveness, it’s hard not to conclude that as we adapt to the intellectual environment of the Net our thinking becomes shallower” (Carr 2010, xx).

In short, the price of the Internet is a less reflective thought process and as such, a shallower life. This is a conclusion similar to Heidegger’s in regards to the price that technology exacts on our ability to reflect and meditate, and to be human. Carr makes an argument against the positive consequences of the Internet by focusing on the detrimental ramifications that he fears are being overlooked by technophiles. He observes that owing to protracted technological usage, his mind was no longer processing data as it once did, including his ability to read books. Similar arguments are being made in relation to human memory and the extent to which it has been externally placed in the Internet (Sparrow et al. 2011). Carr argues that websites tax cognition and diminishes comprehension.

DOI: 10.1057/9781137312402

He states that the cerebral atmosphere of the Internet is "like trying to read a book while doing a crossword puzzle." This he blames on the overload of the working memory from the information strain of the Internet. Neuroscientists will be studying these effects for years to come; in the meantime, technophobes and technophiles of bygone eras are reemerging, making similar arguments though about different devices. Is there, then, indeed something radically different in the Digital Age as compared to earlier stages of technoscientific development?

The view that digital technologies are more invasive and have a greater impact on us than previous technologies have had seems pervasive. Evgeny Morozov, for one, argues that there are two foremost delusions concerning Internet usage, both associated with the personal and social benefits of the Internet as a source of knowledge. For him, "cyber-utopianism" (the belief that the Internet generates freedom) and "internet-centrism" (the idea that every query about the social order and political structure can be based on Internet findings) are two misconceptions that are equally harmful, though for different reasons. We intuitively appreciate the counterarguments regarding the promises of freedom as they relate to the Internet as well as those associated with the overly focused attention we pay to data emanating from websites on the Internet (Morozov 2011). This concern is detailed by Eli Pariser (2011) as well when examining the perils of personalization, especially with regards to search engines (such as Google, Ask, or Yahoo), and how they control what information each person receives based on past searches. One reason Pariser gives for its detriment is the fact that personalization is possible only because websites can collect vast amounts of data about their users, generating a Big Brother-like screening system – a confidentiality catastrophe with indirect control mechanisms.

Though there are many detractors of the digital culture, its environmental hazards, and the particular ways in which we may be personally and collectively hurt by it, there are many supporters and promoters who find a great deal worthy of praise. Clay Shirky (2010) believes in the ever-increasing democratization of information and the great benefits that are bound to follow. For him, there is a direct evolutionary line, even if slow at times, that our story is one of ongoing historical progress. The Gutenberg printing press brought about the Reformation, which heralded the scientific revolutions, which brought about the Enlightenment, which eventually brought to us the Internet. The Internet itself will bring about greater prosperity and human happiness almost in the same sense

DOI: 10.1057/9781137312402

that the French philosopher Antoine-Nicolas de Condorcet described human perfectibility in the Enlightenment (1979/1795).

According to Adam Gopnik, there is another way to approach the literature of the Internet and the Digital Age in terms of three groups: the "Never-Betters" who feel that the Web could always be enhanced by innovative technology; the "Better-Nevers" who wish the Internet had never come into existence; and the "Ever-Wasers" who believe that the Internet is simply our generation's version of disturbing progress/change. He likens the rise of the Internet to the conception of Gutenberg's printing press, and the Protestant Reformation that followed. Just like the aforementioned example, so too is the Internet a technological advance with religious, philosophical, and sociological ramifications. He states, "If ideas of democracy and freedom emerged at the end of the printing-press era, it wasn't by some technological logic but because of parallel inventions" (Gopnik 2011, 125).

No matter what classification one adopts, no matter what binary oppositions one outlines, it becomes clear from this survey that the concern with the onslaught of digital technologies isn't as deep as it is widespread, so that in fact technoscience as such cannot be singled out as the cause of all evil: socioeconomic and political frameworks contribute to or undermine the positive consequences available through these technologies. It is also clear that the postmodern emphasis on displacement rather than replacement holds true: contemporary forms of technoscientific advances find themselves alongside older forms of expression and communication and not necessarily fully supplanting them. That is, we still teach students how to think critically, whether online or in classrooms, assign readings (no matter the method of access), and get paid for our work (no matter the location).

When postcapitalism is introduced as an organizing economic framework, assumptions about abundance may accompany those about scarcity so as to transform the argument about a political economy where people count and prosperity is shared by all. Incidentally, this view is echoed in Brink Lindsey (2007) where he suggests that "post-scarcity" and "postmaterial values" are at the heart of contemporary American culture and politics. Though not focused on this, he'd agree that digital technologies in general and the digital revolution in particular are touted as the means by which abundance can once and for all be achieved. But, as we are experiencing one financial debacle after another, these material conditions alone cannot bring about a radical political economic

DOI: 10.1057/9781137312402

transformation without a radical change in attitude about competition and cooperation, the individual and the community, rights and duties, and an overall reconfiguration of the role of governmental institutions as guardians and mentors rather than obstacles and scrooges. Not only isn't it "one size fits all" when it comes to digital technologies, it's neither a "silver bullet" that can solve once and for all our troubles. Like any other set of innovations, digital ones have to be treated carefully and critically by users and observers alike. With this in mind, we may avoid some of their pitfalls and enjoy their fruits. We are, as ancient myths remind us, playing with fire!

DOI: 10.1057/9781137312402

2

Postcapitalism: Materialism, Decentralization, and Globalization

Abstract: *Focusing on the material conditions of our existence and the evolutionary trajectory we have followed, we can appreciate first that capitalism and socialism have much in common, hence postcapitalism, and second, that group survival is just as powerful a mechanism for natural selection as the so-called survival of the fittest. Digital technologies seem to foster decentralization but paradoxically permit greater concentration of resources and power in the hands of the few, while encouraging a culture of distraction. The implication for the global proliferation of these technologies is greater wealth but one that cannot overcome global pockets of poverty.*

Sassower, Raphael. *Digital Exposure: Postmodern Postcapitalism*. Basingstoke: Palgrave Macmillan, 2013.
DOI: 10.1057/9781137312402.

I A materialist approach

From marketing to sales, from access to financial data to online trading, one of the central pitches for the proliferation of digital technologies has been the idea that only they can revitalize the ideals of capitalism. Adam Smith's idealized marketplace can now flourish the way it never could before: allowing any individual access to any market (equal opportunity and freedom). Although this may be true in the competition for some goods and services, its reverse is true as well: digital technologies allow for the centralization of economic power (large financial institutions: investment banks and hedge-funds). The illusion of decentralization and individual power evaporates once larger entities gain monopolistic power. A materialist critique only adds to what Smith admitted to in his treatise on moral sentiments: a moral social setting is prerequisite for a smooth-functioning marketplace. Against the backdrop of a hybrid postcapitalist economy, the follies and gifts of the proliferation of digital technologies will be examined: do they inevitably lead to growth and abundance? Do they accelerate the rate of wealth concentration, à la Marx, and eventual crises? This also means a gesture toward globalism as an extension of hyper-capitalist practices rather than a greater appreciation of economic solidarity and mutual assistance. Is micro-financing in developing economies digitally inspired? Is it instead a rethinking of the role of women in the maintenance of household economies?

Revisiting the binary outlined in the previous chapter (Table 1.2) in regards to the differences between capitalism and socialism, it's helpful now to introduce more fully what a hybrid vision of postcapitalism looks like (Table 2.1).

Postcapitalism is a collaborative effort that incorporates competition, sometimes called motivation or incentive, for success. In accepting the importance of providing incentives, postcapitalism recognizes national, group, and corporate effort, thereby redefining the notions of rugged individualism and entrepreneurship in more nuanced ways. With this dialectical approach, one can appreciate the institutional role government can play in the affairs of individuals. In extreme cases, lack of coordination can bring about disastrous results. A frequently cited case is the 1904 Baltimore fire that claimed 1,500 buildings and lasted thirty hours. Neighboring cities and states tried to help but because the fittings of their trucks' hoses didn't match those of Baltimore's fire hydrants their services were useless. Sometimes regulatory bodies that ensure universal

DOI: 10.1057/9781137312402

TABLE 2.1 *Capitalism, Socialism, and Postcapitalism*

CAPITALISM	SOCIALISM	POSTCAPITALISM
Economics	Political Economy	Political Economy
Laissez-Faire	Planning (five-year plans)	Information Planning
Competition	Cooperation	Strategic Collaboration
Money (only measure)	Honor/Dignity	Civic Responsibilities
Individual Self-Interest	Collective Consciousness	Sympathy/Benevolence
Division of Labor	Prescribed Duties	Collaboration
Supply and Demand	Central Planning	Regulation/Subsidies
Market Efficiency	Inefficiencies	Disequilibrium
Transparency/Trust	Secrecy/Manipulation	Exposure/Disclosure
Risk (calculated)	State Protection	Regulated Uncertainty
Freedom/Equality	Power Relations	Mediated Freedom/Equality
Private Property	State-owned Property	Private–Public Partnership

fittings can improve the safety of a city, just as may be the case with universal hookups for laptop computers or cell phones.

Even from the depth of the American capitalist literature, associated with management gurus, the terms socialism and capitalism were acknowledged to deserve a reconceptualization at the end of the twentieth century. Peter Drucker, a management expert and columnist for the *Wall Street Journal*, concedes that "if socialism is defined, as Marx defined it, as ownership of the means of production by the employees, then the United States has become the most 'socialist' country around – while remaining the most 'capitalist' one as well" (Drucker 1993, 6). As he notes, pension funds of large corporations regularly invest in the stock market, owning millions of shares of corporations that own the means of production. Century-old definitions of ownership need redefining. Though he liberally uses the term "post-capitalist" in a historical sense of a new society that is heavily based on the knowledge industry, his own definition suggests a "system in which transnational, regional, nation-state, and local, even tribal, structures compete and co-exist" (Ibid., 4). He is on the mark in conceiving of multiple systems simultaneously at work to remake the marketplace; but he's off in envisioning anything more than a management revolution that must account for different priorities. Just a few decades later, radical transformations have ensued, requiring a deeper recognition of human traits and their expression in a

DOI: 10.1057/9781137312402

postcapitalist environment as well as the ways in which digital technologies are prone to crises that invite regulation.

A deeper argument should be made about the nature of our social interactions, one that depends on interpretations of scientific data collection. However one interprets Charles Darwin's evolution theory of 1859 (and Alfred Russell Wallace's), from mutations and variations interacting with different environments and the ability to survive and perpetuate certain traits rather than others, it's not necessary and perhaps inappropriate to adopt the social-Darwinists' view. According to Richard Hofstadter, the notion of the survival of the fittest buttressed a conservative worldview that justified both competition among individuals as a natural phenomenon and considered any victory or success as natural (rather than a matter of luck). This view also neutralizes nature on the path human evolution takes so that any forced reforms or tinkering (government intervention) are deemed contrary to natural laws (Hofstadter 1955/1944, 6–7). This particular interpretation bolsters a particular ideology that supported a social and economic view of American culture. Under natural conditions of competition, it is natural and therefore morally acceptable that some would survive and thrive while others perish; any intervention runs afoul of nature and by extension God's design.

What fascinates Edward O. Wilson, by contrast, is that even when accepting the general outlines of Darwin's evolution theory, as he does, the unit of selection may not necessarily be the individual organism. And when this comes into question, a different view emerges of human social behavior. In order to appreciate his view of "multilevel evolution," it's helpful to follow some steps in his argument. In Wilson's words: "One of the principles is the distinction between the unit of heredity, as opposed to the target of selection in the process that drives evolution. The *unit* is a gene or arrangement of genes that form part of the hereditary code ... The *target* of selection is the trait or combination of traits encoded by the units of heredity and favored or disfavored by the environment. Examples of targets are propensity for hypertension and resistance to disease in humans, or, in the case of bird behavior, the instinctive choice of nest site" (2012, 162, emphases in original).

Genes are the hereditary carriers of units and not individual people, just as targets of selection are a set of traits that become apparent in individual humans. Propensities themselves are not determinants, since environmental factors are the final arbiters of actual results, such as the presence of a disease. Wilson continues: "Natural selection is usually *multilevel:* it acts on

DOI: 10.1057/9781137312402

genes that prescribe targets at more than one level of biological organization, such as cell and organism, or organism and colony" (Ibid., emphasis in original). Multilevel evolutionary trends are evident, according to Wilson, in the behavior of certain insects that find themselves in colonies fighting off intrusion by other colonies. What he calls "eusociality" is defined as "group members containing multiple generations and prone to perform altruistic acts as part of their division of labor" (Ibid., 16). "Individual-versus-group selection results in a mix of altruism and selfishness" (Ibid., 163) insofar as "selection among genetically diverse individual members promotes selfish behavior ... selection between groups of humans typically promotes altruism among members of the colony" (Ibid., 162).

With this in mind, Wilson concludes, "Our species is not *Homo oeconomicus*. At the end of the day, it emerges as something more complicated and interesting. We are *Homo sapiens*, imperfect beings, soldiering on with conflicted impulses through an unpredictable, implacably threatening world, doing our best with what we have" (Ibid., 251). Wilson draws his conclusions from studying ants first and humans second, but when he emphasizes multilevel selection and evolution, it provides him a basis from which to distinguish between our tendencies to be selfish and cruel on the one hand, and benevolent and altruistic on the other. Recognizing two sets of motivation as co-determinants of human evolution robs the social-Darwinists the foundation of their view. Social – and by extension economic and moral – behavior is inherently conflicted and complex, leading some communities to fall apart while allowing others to cooperate and thrive. This is true of the feedback individuals receive from the consequences of their behavior, just like some colonies that cooperate and survive for thousands of years as opposed to those that fall prey and succumb to predators quite easily.

As for postcapitalism, it's clear why a Wilsonian interpretation of natural selection is more informative than a social-Darwinian one. While the former allows for multiple modes of behavior depending on contexts (worthwhile to cooperate, dangerous to collaborate), the latter expects "one size fits all" – no matter under what conditions, humans are bound to compete for survival (even when survival, however defined, isn't an issue at all). This view is echoed most forcefully by Martin Nowak who studies mathematical models of cooperation based on the Prisoner's Dilemma. Just like Wilson, he is concerned with the conditions under which collaboration not only happens in nature and therefore should be emulated, but also the necessity of such human collaboration to avoid

DOI: 10.1057/9781137312402

the "tragedy of the commons" where individuals abuse common goods, such as land and air. For him, there are five mechanisms that should be consciously added to the debate about human behavior and interaction in terms of a sustainable evolutionary path: repetition, reputation, spatial selection, multilevel selection, and kin selection (Nowak 2011, 270–272). These mechanisms are to be added where missing or highlighted where undervalued so that people would be more collaborative as role models to others, and as guardians of the commons that benefit everyone. Following others, Nowak invokes the hope that Adam Smith's "spirit" should be "exorcised" so as to remove the appeal of the "invisible hand" and replace it with cooperative mechanism (Ibid., 204). For Nowak, then, the issue isn't competition *versus* cooperation, but that "cooperation and competition are forever entwined in a tight embrace" (Ibid., 269).

Contextualizing Wilson's theory of multilevel evolution and Nowak's ideas about cooperation that could support a postcapitalist view with reasonable occasions for government intervention (incentives to pursue certain paths rather than others, as in the case of energy production) or regulation (safety issues, from fire codes to airport controls), it's helpful to consider our digital connectivity. There seems to be a different kind of urgency to reconsider inviting government agencies to meddle in our lives, not as in the case of Big-Brother-like already covered by Nowak's interest in networks and their influence on behavior (Ibid., ch. 12), but rather as in the cases of digital intrusion into every facet of our regular lives, such as shopping and eating, working and driving. According to William Davidow, contemporary society needs to be more fully aware of the dangers of its "connectivity" – the manner in which every transaction and institution is linked through the Internet to every other individual and institution. Following his outline, "when it comes to degrees of connectivity, there are four general classifications that can be applied to nations, economic regions, and societies:

1 Underconnected state. Isolated ancient civilizations, primitive cultures, and undeveloped countries are examples of underconnected states ... [Iceland before the twentieth century]
2 Interconnected state. In this state, when the environment changes gradually, businesses, economic systems, and governments are capable of keeping pace ... [Chicago before railroads]
3 Highly connected state ... In this state, businesses, economic systems, and government institutions are driving change ... society can still cope and prosper. [Silicon Valley in the late twentieth century]

DOI: 10.1057/9781137312402

4 Overconnected state. In this state, institutions change so quickly that the environment in which they are embedded is unable to cope. [Iceland's recent financial crisis]" (Davidow 2011, 21–22)

What's insightful about this four-stage genealogy is the focus on different kinds of connections, from telegraphs and railroads all the way to the Internet, and the material conditions they provide for certain human and social interactions. Those in turn are either in or out of sync with and correlate to other modes of connections and relations – railroads and trade and banking and labor mobility. When out of sync, when one set of material conditions lags behind another, when the infrastructure, for example, of roads and railroads cannot accommodate trade transit and the shift from agrarian rural society to an industrial urban one, then government institutions cannot cope with discontent or even strife, such as not enough lodging for large numbers of migrants or roads too narrow for large trucks. As "the success of Silicon Valley was driven by hundreds of positive feedback loops," (Ibid., 33) it was also helped by local and national government agencies that kept up with appropriate regulation in an accelerated rate of connectivity, so as to match connectivity and ensure prosperity (Ibid., 41–42). The keeping up in this case was not limited to roads and power grids, abundant buildings and infrastructure, but included regulatory frameworks to encourage the development and protection of intellectual property (more on this in Chapter 4). Once the shift is from the materiality of government control (roads, power-lines, etc.) to a virtual space where connectivity is tested and advanced, a new set of problems arise. According to Davidow, "As interconnection technologies have extended the reach of business and helped it migrate from physical to virtual space, the bonds between the pursuit of self-interest and the social control Smith envisioned have been further weakened. To see how this is happening, we need merely compare Smith's time to our own" (Ibid., 185–186).

Lamenting the bygone era of small villages in Smith's time is unproductive. Instead, we may have some control over the speed with which we move from one connectivity stage to another and attempt to slow the process, or even reverse some of its extreme manifestations: "instead of an overconnected world we return to a highly connected one... need to focus more attention on prevention... introduce controls... Regulation is one important way to reduce risk to depositors... to the economy as a whole" (Ibid., 189–192). Obviously increased controls may have the unintended consequences of undermining progress in a digital culture, even

DOI: 10.1057/9781137312402

paralyzing potential developments. But it's one thing to recommend that private institutions not become too large to manage or overly connected, and quite another to set limits on the natural growth of private and public institutions. Perhaps here is where incremental changes both to digital technologies' development and to government regulation might be useful (Popper 1957).

II Decentralization/centralization

For Adam Smith the world of commerce functions best when all its members, whether within the village marketplace or in the international arena of trade, remain small enough not to affect the market as a whole. In this sense, decentralization means the assurance that power constellations do not form to gain unfair advantage when trading. He was worried about collusion and the conditions under which its impact is minimized, appealing to regulation (Smith 1937/1776, 691–696). Karl Marx was even more concerned with monopoly power, and predicted an ever-increasing concentration of capital to the detriment of all those who remained outside its orbit. For him, the inevitable, historically determined accumulation of wealth would lead to crises that eventually bring about the deterioration and disintegration of the capitalist system as a whole (Marx 1976/1867, 80–86). Instead of following their respective arguments, we should perceive the current conditions of the economy so as to appreciate, at the very least, the paradoxical situation in which there is a great deal of concentration of wealth and capital in the hands of the "1 percent" (in the US), while claims for decentralization keep sounding all the way from academic quarters (as we shall see below) as well as apologists who remind us all that anyone can be a millionaire with a lot of hard work and a bit of luck.

As far as Yochai Benkler is concerned, "the Digital Age and its attendant Internet is the first communication medium that expands its reach by decentralizing the capital structure of production and distribution of information, culture, and knowledge ... The basic change in the material conditions of information and cultural production and distribution have substantial effects on how we come to know the world we occupy and the alternative courses of action open to us as individuals and as social actors. Through these effects, the emerging networked environment structures how we perceive and pursue core values in modern liberal

DOI: 10.1057/9781137312402

societies" (2006, 30). Though not intending to fall into the Marxist column of capitalist critique or capitalist apologetics with a conscience, Benkler displays an uncanny allegiance to a materialist view of cultural development insofar as he admits that the material conditions of digital technologies (production and distribution) determine our society in terms of values or superstructure (social conventions, moral norms, and legal frameworks).

For Benkler, there are two phases in the shift to the information age: "the industrial information economy" where economies of scale (the more is produced the cheaper the product per unit becomes) "led to the rise of the concentrated, controlled form of media," and then "the networked information economy" that offers "many of the diverse paths and mechanisms for cultural transmission..." (Ibid., 32). This second stage of development becomes possible because material conditions have radically changed so that a company or municipality, for example, doesn't need to lay cables anymore, and every start-up company or group of people can ride the Internet wave and exploit an already existing infrastructure at minimal cost. Barriers to entry, as the common technical term suggests, are so low that almost anyone with a laptop and Internet connection can with minimal effort sell whatever idea or object to anyone else in the global marketplace. Services, such as eBay, provide relatively secure and reliable platforms for market-like exchanges across the nation and internationally, providing third-party payment methods that ensure the completion of individual transactions.

Benkler put in motion a whole new appreciation of what peer production can bring about without planning by either government agencies or private corporations. As Steven Johnson picks up this thread of thought, relying on data from the past two decades, it becomes apparent that even mega-efforts, such as the Internet have neither been government-sponsored efforts nor the exclusive investment of some giant global corporations. Instead, he frames his observation about networks of peers in terms of the general assumptions we have about large undertakings, what has been called Big Science in the US: it's either a capitalist effort or a socialist one, either an immense and risky investment by a single visionary CEO of a wealthy company or a government underwriting through Department of Defense appropriation. What is lost in this article of faith is the simple fact that, as Benkler shows (2006, ch. 3), material conditions related to digital technologies actually allowed peer networks to thrive, no differently from the ways in which they have successfully brought

DOI: 10.1057/9781137312402

about the scientific revolutions and the Enlightenment of the sixteenth through the eighteenth centuries (Johnson 2012). What is fascinating about this narrative of decentralized efforts that lead to an entirely novel transmission of news and ideas, entertainment, and market exchanges is that it recognizes the simultaneous need for some governmental indirect support – laws and rules – as well as entrepreneur initiatives – individual commitments and risk-taking, all blended together in a hybrid economic environment whose destiny no individual or institution can dictate or dream up in advance. But this narrative has some unintended negative consequences we ought to recount.

Digital technologies pretend to empower individuals and thereby diffuse power concentration in the hands of the few, as Benkler claims. But, as we have seen more recently, the tendency to merge companies or acquire them remains strong in the corporate world, based on the belief that economies of scale are always at hand even when integration becomes expensive and problematic when different corporate cultures cannot find a common language or ethos around which to combine their respective efforts. With the advent of Constitutional permission for unlimited expenditure on political causes (*Citizens United v Federal Election Commission, 2010*), economic collusion now has political protection from regulatory intervention. As corporate giants, such as Goggle and Facebook, Apple and Microsoft, keep increasing their market share and dominance, the ideals of decentralization may remain mythological in nature. By the end of 2006, there were eight "giant media companies" controlling news and information: General Electric-NBC ($390.6 billion), Microsoft ($306.8 billion), Google ($154.6 billion), AOL-Time Warner ($90.7 billion), Disney ($72.8 billion), News Corporation ($56.7 billion), Viacom ($53.9 billion), and Yahoo! ($40.1 billion) (*Mother Jones*). This comes close to what one sees in other sectors of the economy, such as agriculture, where giants like Monsanto control almost 90 percent of some genetically modified seeds.

It could be argued that mega-corporations in the information industry provide cheaper product – at times even free – to consumers, since advertisers end up paying for distribution channels. Likewise, one could argue that mega-media corporations are not beholden to any political regime and therefore transcend local politics. This allows them to bring more objective news to states where governmental control limits the range of news options the citizens have. Being trans-political, or at least not government-bound, allows the likes of Google to remain above the

DOI: 10.1057/9781137312402

fray. Media giants become resources for dissidents to tap into despite restrictions in ways that previous technology could hardly match (unless one thinks of short-wave radio transmissions as equivalent in principle, but of course not in magnitude). In this context we should note that youth empowerment is readily available and can coalesce into blogging groups, such as Anonymous, whose impact on the formation of Occupy Wall Street in 2011–2012 is noteworthy. Unlike corporate giants, of course, individual bloggers don't have the same political impact, and can be easily shut down.

As the Arab Spring of 2011 illustrated, the decentralization of digital technologies has its limits insofar as they can be curtailed if not completely undermined. When the Egyptian authorities felt public outrage was too vocally expressed through digital devices, they shut down the Internet within its borders. Syria has recently followed suit. Similar actions are regularly undertaken in China when the Communist Party leadership finds it prudent to do so. And more recently, the Iranian government announced in the fall of 2012 that it's developing an "alternative Internet" where websites are sanctioned for public use. In all of these cases, those with digital capital – knowledge of how to circumnavigate governmental controls – may find ways to retain their independence and exchange information across barriers and filters; but they remain the minority of digitally sophisticated users, while the vast majority remains bound by the restrictions set up by governmental agencies. Incidentally, this kind of government behavior with regards to the Internet and digital devices is not the exclusive domain of dictatorships and single-party nation-states; rather, most democracies resort to such behavior under the banner of "national security." Censorship is routine, and surveillance of public behavior on the Internet is quite common. The issue here is no longer the concentration of the production of the infrastructure of the information economy, but its distribution. Even when large corporations control large segments of this economy, government agencies have been able to collect personal records of citizens or directly monitor Internet behavior. One could argue that unlike the old-fashioned internal spying the Soviet Union undertook of its citizens, the US today can much more easily spy on its own citizens by setting up broad-ranging warning devices that alert it to any information exchange with "undesired" sources. The infamous *Patriot Act* of 2001 gave various governmental agencies the authority to monitor communication without any case-by-case legal sanction (warrant). The Digital Age has enhanced this kind

DOI: 10.1057/9781137312402

of concentration of powers despite the presumption that because each of us has an encrypted account or smartphone we are therefore more autonomous. It's true that we have access to more channels of information than ever before, but all of them can be accidentally or deliberately shut down instantly.

Of relevance here is the extent to which the Commons remains an important element in the decentralized, individuated digital culture as we have seen earlier in Benkler's phrasing of "commons-based peer production" (2006, 60), some of which relates to disputes over intellectual property rights (see Chapter 4). Despite the great concentration of digital expertise and wealth in the hands of the few, there must be some place – physically, cognitively, and virtually – where corporate giants meet individual consumers. Traditionally, there was the public square where people from all walks of life encountered each other. No matter one's occupation or socioeconomic status, there were occasions when people crossed paths and realized that they had enough in common to warrant the label of a citizen of this or that village or state. With this in mind, it would also seem reasonable to propound the principles under which a social contract was implicitly agreed upon by everyone: each has rights and duties that are sanctioned by the general will that encompasses them all. Military service comes out of this thinking, whether mandatory or voluntary, as well as all those public goods neoclassical economists talk about (parks, roads, bridges).

In an affluent age when commercial flights are segmented to First Class, Business Class, and Economy, or where gated communities have their own security guards, the very idea of a common ground shared by all seems more far-fetched. Put differently, an economic arrangement that encourages class stratification based on the accumulation of wealth is bound to leave pockets of poverty that may or may not be attended to by a progressive tax-system. The Digital Age may have enhanced this development rather than retarded it; it may have contributed to the general distraction of the population from crucial economic trends that can be mitigated with policies and regulation. Perhaps it was television culture that hooked millions to a screen for hours every day as a replacement for ancient public amusements, such as gladiators in the arena or an occasional beheading; perhaps it was modern consumer-culture that allowed every household to buy a television set and a video machine for individual entertainment, separate from the crowds of a movie theater. Perhaps it is this *culture of distraction* that ensures we are glued

DOI: 10.1057/9781137312402

to our laptop screens or smartphones, playing games for hours rather than exchanging ideas in public forums (except in blogs). The culture of distraction moves seamlessly along a digital trajectory that separates us from each other, that allows us to be self-contained entertainment units far away from any constellation of concern or complaint. *For us, then, economies of scale have brought the riches of gadgets we feel compelled to consume, feeling highly connected with the world while retaining our distance from the material reality of the public square.*

The culture of distraction is highly addicted to digital devices because of their ease of use and inexpensive price tag. No longer tied to the radio or television console in one's home, smartphones transmit the world's news literally into one's hand. The mobility of these devices compounds our distraction because they never leave us, whether we walk, jog, drive, or fly. Instant connectivity also ensures a constant barrage of advertisements, an incessant heap of temptations we can consume at the click of a finger. We no longer need to shop in stores or malls; we no longer need to encounter other people during this process of acquisition. This is true of all consumable goods and services including, sadly enough, education as online courses are readily offered by the best-known universities free of charge.

III Globalization/poverty

Smith conceived of his world as made up of nation-states that have trade agreements and tariffs to ensure the kind of trading or exchange he envisioned in the village marketplace. His extrapolation from village mentality to that of the world has been tested over the centuries, and has emerged again as, for example, the US complains about unfair trade practices of China. The Chinese are said to regularly subsidize their export industries so that they have an unfair advantage when trading in the global economy (not to mention manipulating the value of their currency). These concerns have not left us. Some critics, such as Joseph Stiglitz, suggest scathing critiques of the International Monetary Fund, contending that the IMF has exacerbated the financial situation of less developed countries which is even worse than not intervening on their behalf at all. According to Stiglitz, globalization as a notion that eliminates barriers to trade in Smith's sense can improve the global economy

DOI: 10.1057/9781137312402

or can make it worse (2003, Preface). Smith's ideal of peaceful free trade that improves the well-being of all participants remains alive today.

Likewise, Marx's perception of the economy as a battleground between capitalists and proletariat has retained its intellectual and practical bite. When Marx invited the workers of the world to unite, he envisioned the reality of outsourcing like no one before his time. Corporate giants and small businesses alike look for the cheapest labor sources around the world so as to escape escalating labor costs at home. But global laborers have not united, and the cost differentiation has made some countries, such as India and China, Singapore and Taiwan, more appealing to manufacturers as compared with the US and the European Union. The Digital Age has not overcome the material conditions of this battleground but in fact made them more apparent insofar as physical limitations have been digitally transcended, sending plans, designs, and models electronically across the globe for low-cost bids. Whether we are talking about call-centers located across the oceans or manufacturing facilities, the global economy looks very much like Smith's village marketplace where anyone can participate even though nowadays it begins virtually before it becomes a physical presence.

While Smith and Marx in their respective ways still uphold an idealized capitalist model of the economy as the standard against which to offer their proposals – either for a free and equal marketplace or labor battlefield for wealth and power – Jeffrey Sachs suggests that this model is no longer applicable in the late twentieth and early twenty-first centuries. For him, the global economy now "shares a *common fate on a crowded planet*" (2008, 3, emphasis in original). This means that any trade advantage is temporary at best and every crisis isn't containable in any one region of the world. Instead, as was proven time and again in the past few decades, every national crisis – from mortgage financial instruments to droughts – becomes international quite rapidly. As Davidow argues, ours is an overconnected world where any economic factor affects every other, where any financial movement in one market is affecting trades around the globe. It is under these conditions of digital connectivity that Sachs insists on cooperation among the global economies, admitting along the way that they are indeed united in some fundamental ways. Without using the terminology of postcapitalism as outlined above, Sachs in fact promotes a hybrid approach that mixes the positive effects of competitiveness in terms of efficiency as well as those of coordination among competing interests so as to ensure a level of sustainability (rather

DOI: 10.1057/9781137312402

than ongoing crises). According to this global model, there must be a more nuanced and enlightened view of how a united global economy (giant corporations and international markets for everything from oil to seeds) operates within and among divided social and political entities (nation-states).

As we saw with the latest financial crisis of 2007–2011 (and its ongoing impact) labeled fondly the Great Recession, its presence is felt from Main Street US all the way to the European Union and other mega-markets in India and China. When banks in America stop lending money to small businesses, products made in China are not purchased; when Chinese workers demand higher wages, Wal-Mart customers feel price increases; when the Greek economy collapses, when Spain experiences 25 percent unemployment, international banks are worried about the risk associated with their respective bonds and their interest rates. Iceland was only one concrete and extreme example when its banks speculated with others around the world and brought their small nation to an economic crisis in 2010–2011. When teachers' pension funds invest in credit derivatives the origins of which they do not know, it's not surprising that they might lose their members' savings. Economic factors cannot be that divorced from the people who are directly and indirectly engaged with them. It's this relationship between the economy and the social and political frameworks within which it operates that must be emphasized – an emphasis not lost on either Smith (see his treatise on moral sentiments of 1759) or Marx (see his entire life's work where political structures are analyzed).

Though developmental economists, like Sachs, appreciate the need for coordination in a networked global economy, they remain fairly timid, on ideological grounds, suggesting holistic solutions (Sachs 2008, ch. 14), perhaps because they reek of some centralized planning. The ghosts of the Soviet Union are still with us, and when they are coupled with a belief in some unplanned peer coordination (Benkler 2006), this leads to complete distrust in any coordinated planning. In the period between the two world wars of the twentieth century, Paul Rosenstein-Rodan (1961) proposed his novel idea of the "big push" – a coordinated effort to provide an infrastructure that would accommodate individual efforts to develop this or that industry, this or that water project or farming. This, incidentally, was well before digital technologies enhanced our capacity to know what's happening around the globe as well as allow those around the globe to have access to the mysteries of the decisions

DOI: 10.1057/9781137312402

undertaken by central banks and wealthy donor-states. Rosenstein-Rodan's ideas follow the Smith–Marx axis of materiality, but consider an institutionally peaceful approach, however revolutionary, rather than a call to arms, as Marx's approach has been historically interpreted.

As Sachs agrees almost a century later, case-by-case business investment, philanthropic or self-interested, may fail if it isn't part of a holistic approach to ensure sustainable development. One can look at the recent foreign-aid experiment in Afghanistan, a decade-long foreign investment that despite or perhaps because of digital accessibility has yielded no results for sustainable autonomy to a fractured country beseeched by tribal strife and a thriving drug exporting industry. Perhaps the lessons of such failures have to do with our misguided approach that refuses on ideological grounds to appeal to a full-fledged coordinated planning that is not limited to self-policing military stability.

This holistic approach also accounts for the inherent disadvantages some countries have in comparison to others, with an eye to eliminating as many as possible pockets of poverty. Global discrepancies may be justified by those who have an ideological commitment to social-Darwinism (where the unit of analysis is a state), judging poor economies in terms of personal laziness and institutional corruption. Perhaps because of this reason some, like Scott Barrett (2007), offer a view of cooperation that depends on coordination among nations, acknowledging the fact that some countries just cannot get their own domestic policies under control to fulfill any global obligations for any international cooperation. Their discussion of the global situation should be understood against the backdrop of Wilson's multilevel evolutionary process, where there is an advantage for individuals to pull resources together in a group in order to survive. What if we began to perceive our "crowded planet," as Sachs suggests, as one very large group in Wilson's sense, whose survival can be ensured only if all its members collaborate? And how would we protect cultural diversity from the ease with which cultural imperialism is enhanced by digital technologies?

DOI: 10.1057/9781137312402

3
Political Economy: Freedom, Surveillance, and Entrepreneurship

Abstract: *The legacy of the Austrian School relates to the separation of the state from the marketplace so as to ensure maximum individual freedom. Supercapitalism's ascent coupled with the advent of digital technologies enabled a new level of surveillance in the workplace. The customized surveillance of consumers' tastes and preferences seems friendly but is in fact controlling. Perhaps the entrepreneurial spirit associated with postcapitalist crowd-funding allows a modicum of personal liberty envisioned by the Austrians.*

Sassower, Raphael. *Digital Exposure: Postmodern Postcapitalism*. Basingstoke: Palgrave Macmillan, 2013.
DOI: 10.1057/9781137312402.

I Freedom

This chapter focuses on the political implications of hybrid economies in the Digital Age. It will review some of the standard arguments that align new digital devices with the potential to express freedom and bring about revolutions, while acknowledging counterarguments that suggest that technoscience alone cannot bring about freedom. This is an empirical as well as philosophical argument. The subtext of this debate is the underlying concern with the concentration of political power of surveillance that has been enhanced over time (see the Patriot Act in the US after September 11, 2001, Egypt right at the onset of the revolution in 2011, and Russia around the 2012 presidential election). So long as the political framing of any economic system is essential for promoting certain economic principles over others, what is at stake is old-fashioned political economy as understood by eighteenth and nineteenth century thinkers. Domestic forms of controlling citizens – politically and economically – are apparent from the former Soviet Union satellite states to the democratic US, whether under the rubric of "regulation" or the tax code.

Theoretical concerns are constructed in light of imminent practical threats, crisis-situations, or wars. The hyperbole in the past couple of years of the so-called Arab Spring was in direct reaction to decades-long dictatorial regimes supported by military might, as was seen all the way from Tunisia and Egypt to Libya and Yemen. A similar though different reaction came in the wake of the Russian revolution and the establishment of the United Soviet States of Russia as a quasi-communist state regime as well as Nazi Germany and fascist Italy between the two world wars. The horrors of dictatorships, whether understood as socialist or not, provided the backdrop for the Austrian School of Economics. It should be remembered that its leading members reacted just as much to an earlier debate over the interpretation of history or more specifically the cultural-dependent progression of economies within their national boundaries (also known as the *Methodenstreit* which means "fight over method"). It was the Austrian School which demanded a more "scientific" understanding of the economy so as to divorce political influences from financial markets. As much as the Austrians attempted to quantify their ideas and provide objective standards of measurements – from the psychological principle of utility to that of diminishing marginal returns – they also wanted to ward off what they called "collectivist" tendencies.

DOI: 10.1057/9781137312402

In light of what was said in the previous two chapters about hybrid economies and the integration of capitalist and socialist ideologies, it's worth spending some time revisiting this period of economic thought.

As Ludwig von Mises explains, what was once the Austrian School has in fact become the mainstream of economic thinking, overcoming the dominance of the Historical School. Among the reasons that the Historical School came to disrepute was not only its association with the eventual emergence of Bolshevism, Nazism, and Fascism, but also its insistence that no "'abstract' theory of 'armchair' authors" could account for the different historical conditions each epoch brought about (von Mises 1984, 23). For them, then, government intervention made sense insofar as it could respond to these particular conditions to improve the economic conditions of their countries. By contrast, the Austrian School economists, starting with Carl Menger in the nineteenth century, believed in marketplace arrangements akin to those advocated already by Smith: "As human nature is, it is unavoidable in any society where division of labor and its corollary, market exchange, have reached such an intensity that everybody's subsistence depends on other people's conduct ... The services are rendered voluntarily ... The whole system is built upon this voluntariness of the services exchanged ... But this integration into the community of the market economy is spontaneous, the result of the insight that there is no better or, for that matter, no other method of survival open to him" (Ibid., 42–43).

We move from a revised presentation of Smith's rationale for marketplace exchange to a view of human voluntarism that is Wilsonian in its concerns with the survival of the species, not simply that of the individual. This, of course, sets this model as a natural model based on human nature. But in order to set voluntarism within political institutions of the state, the Austrians also advocated a liberal ideology: "The true liberals firmly held that the market economy, the only economic system that warrants a steadily progressing improvement of mankind's material welfare, can work only in an atmosphere of undisturbed peace. They advocated government by the people's elected representatives because they took it for granted that only this system will lastingly preserve peace both in domestic and in foreign affairs" (Ibid., 35–36). *For the Austrians, economic freedom goes hand in hand with political freedom; the one cannot be accomplished without the other.* Which one takes precedence in any historical or logical sense becomes irrelevant or even reductionist in the wrong sense of a positivist attempt to find a cause and effect

DOI: 10.1057/9781137312402

relation between the two. Instead, it seems that to respond to nature's call of what humans are like we must adopt a liberal ideology that guarantees freedom of thought and action to all, and with this in mind "spontaneous" marketplace exchanges are bound to appear everywhere, from local markets to international trade. The Digital Age, according to this view, is no different from prior or future ages, and it, too, requires a liberal mind-set of individual freedom that inevitably brings about prosperity for all.

But, as Friedrich Hayek so eloquently warned in the previous century, "Individualism has a bad name today [originally published in 1944], and the term has come to be connected with egotism and selfishness" (1972/1944, 14). He contrasts individualism with socialism and collectivism as alternative forms that inform a social organization in which the welfare of society as a whole takes precedence over the individual. Hayek's warnings were about the ways in which capitalism undermined the ability of individuals to strive for and achieve their own success. He understood that "to create conditions in which competition will be as effective as possible" requires "a wide and unquestioned field for state activity" (Ibid., 39). In order to protect individualism one requires state intervention; and state intervention, whether by representative majorities or lobbying efforts of interest groups, inevitably leads to the kind of collectivism Hayek railed against. Unlike the Marxist prediction of the demise of capitalism in favor of socialism because of the proletariat revolution, this one has come about democratically, with majority-rule that displays what John S. Mill warned about as the "tyranny of the majority." The Austrians' influence in the US was not as much linked with a critical backward-looking view of the horrors of Bolshevism, Nazism, and Fascism of the early part of the twentieth century, but rather the forward-looking view of the potential harm democracy is about to inflict on capitalism as they envisioned it, pushing it all the way toward a socialist collectivism that would undermine and choke any potential for entrepreneurship.

It's important to recall the specter of the Austrians not only because of their ongoing influence on the Chicago School of Economics under the tutelage of Milton Friedman whose *Capitalism and Freedom* (1982/1962) reads like a continuation of and a response to Friedrich Hayek's *The Road to Serfdom* (1972/1944), but also because this way of arguing about freedom brings together economic theory and political philosophy. To bring about or secure the conditions for economic prosperity one must

DOI: 10.1057/9781137312402

deal with the political framework within which they will be legally guaranteed. And this vision of freedom linked as it is to economic wealth dominates present conversations about the fruits of the Internet and digital technologies: founders of mega-corporations, from Microsoft (Bill Gates) and Apple (Steve Jobs) to Google (Larry Page and Sergey Brin) and Facebook (Mark Zuckerberg) are measured in terms of the billions they have accumulated. We follow the ebb and flow of their fortunes as if such an observation gives us an insight into their genius and the conditions that made it possible to thrive, rather than a broader and more nuanced examination of socioeconomic and ethno-class stratification that allowed some to excel while leaving many equally talented individuals behind. *If political and personal freedoms are discussed only in economic terms, and if economic terms are discussed only in the absolute terms of (the Austrian view of) human nature, then of course a very particular view emerges: the myth of the genius working alone to achieve greatness.* But is this view both empirically and theoretically defensible?

While Jean-Jacques Rousseau proposed a "social contract" in order to deal with the conundrum that "man is born free, and everywhere he is in chains," the Austrians are suspicious if not downright hostile to the compromises necessary for such a contract to retain its legitimacy and authority (Rousseau 1978/1762). They rightfully find faults with a political system that requires one to relinquish natural freedoms in order to regain them under state-sponsorship. They have witnessed all too painfully collectivist abuses, and remain distrustful of all promises. But what they articulate as an ideal situation of radical individualized freedom remains either too ideal and therefore unrealistic or too dangerous in its absoluteness. With this in mind, we can move from generalized views of freedom to more specialized ones as they refer to the business world in the Digital Age, examining some of the controversies about state or commercialized surveillance.

II Surveillance

When thinking of curtailing liberties, especially personal ones, what comes to mind is forms of censorship and governmental surveillance. As for censorship, the US Supreme Court, as we have shown elsewhere (Sassower and Cicotello 2010, Ch. 4), has come down on the side of the government in most cases citing national security concerns (in combat

DOI: 10.1057/9781137312402

zones). In the present context, the concern is more with surveillance than with censorship. And when it comes to surveillance, the concern isn't with government agencies – though they probably exceed their Constitutional rights in the name of national security after the 9/11 mass-hysteria in the US – but rather with corporate entities. What's alarming in this specific mode of surveillance is that most individuals submit willingly to it without appreciating the unintended consequences associated with it. There are, though, two kinds of commercial surveillance: internal to the business or organization, technologies known as enterprise systems, and external, those targeting consumers.

As for the enterprise systems, they are primarily used, as Simon Head suggests "to standardize and then monitor the entire range of tasks being done by a company's workforce" (2007, 42). I recall about twenty years ago, just after I was divorced and right when the Internet was being tested on campuses, that I was working on campus late into the night, when all of a sudden a message popped up from my chancellor being impressed and congratulating me on working late. I recall my surprise: how were my whereabouts known to someone much higher on the institutional ladder? And then I realized that anyone can monitor what I do and when I do it. I turned off my Internet connection while working on my books after that, realizing that monitoring workers is in fact undertaken more often than we realize, and that this fact alone is worrisome even when the worker is congratulated. The internal monitoring that Head describes may increase efficiency when applied to manufacturing processes or delivery systems, knowing exactly at what stage or position any item is at any given time. This kind of system made Wal-Mart the giant discount retailer it has become in the past few decades; it also has made UPS and FedEx, to mention only two delivery corporations, so successful. But beyond manufacturing and transportation, these enterprise systems have also been applied to white-collar industries, especially because these "automated regimes" are able to "reduce these complex human activities and reasoning to a series of processes and outcomes that can be mapped out and programmed by a computer" (Ibid.). This means that one's productivity, the amount of work one completes in a set time-frame, is measurable and therefore manageable – and no one needs to be around to "spy" on individuals. Instead, a computer will monitor and record, process the data, and evaluate the efficiency associated with the entire enterprise.

Despite a growing sense that Information Technologies and the advent of computerized workplaces necessarily increase productivity,

DOI: 10.1057/9781137312402

Head explains that according to the McKinsey Global Institute, during the heyday of the digital revolution (1995–2000): "US productivity gains were very narrowly focused within six of the economy's sixty sectors" (Ibid., 43). As expected, among the six sectors were large wholesalers and retailers, distributors, and IT companies. Oddly, health care providers, from insurance companies to hospitals, as well as the banking industry, didn't see much bottom-line impact because of this digital revolution. This is not to say that future developments will not enhance productivity in all of the economy's sectors, but only to suggest that a digital revolution in itself doesn't inevitably bring about greater efficiency to corporate America. Additional cultural and policy matters must be addressed in order to promote wider use of digital technologies as well as some radical changes in corporate culture that could fully enjoy the fruits of these technologies, such as flexible workdays, interchangeable task-sharing by co-workers, and a greater division of all tasks and goals to small, delineated, and measurable units that can neatly find their properly weighted place on an excel spreadsheet. The watchful Big Brother of George Orwell's *1984* has been transformed from the shift-supervisor to the surveillance camera and the computerized reporting system, such that every move and action, every thought and exchange becomes part of the fabric of the institution, allowing corporate leaders to punish and reward according to a rationalized system, an objective standard employees must accept both contractually and personally. Deviation is intolerable and differentiation is outside the norm.

Quite contrary to the standardization of the workplace and the erasure of the individual in the production and distribution of goods and services, the consumption arena thrives on setting extremely personalized consumption platforms, emphasizing customized monitoring that fits all tastes and preferences. The advent of digital technologies in capitalist hands (and not only in the hands of government agencies that spy on the activities of citizens in the name of National Security) enhanced *uniformity in the workplace and singularity of consumption*. There are those, like Pariser (2011), decrying the perils of personalization of search engines because the information they produce on behalf of individuals accounts for past searches, and therefore are self-limiting. These websites collect an enormous data set from their so-called clients (who don't register personally or give explicit permission for the collection of these data) under an implicit understanding that free individual use of the website is contingent on the website's ability to keep such records. With the

DOI: 10.1057/9781137312402

proper algorithm and a large enough storing capacity, companies collect valuable information they are free to use either for their own purposes or to sell to the highest bidder. This might be considered, according to Pariser, a form of censorship because of the filters set to collect certain data points and offer others in return, narrowing the options along the way, and determining, however indirectly, what information is available to individual consumers.

There are those, like Ken Auletta, who attempt to give a more balanced and less alarmed presentation of the workings of websites, such as Google. He inspects both viewpoints of the Google privacy deliberation: from the privacy advocate's point of view (Why does Google need to track my search history?) and from Google's point of view (the public's privacy concerns are silly fears that we will do malevolent things!). Though Auletta does not offer his own view, he argues that privacy is one of the several impediments that Google will have to circumvent in the future in order to continue "surfing a huge wave that seems not to have crested" (2009, xx) This probably will never force the Googles of this world to stop collecting personal data, however encrypted, but might curtail their boastful statement about such powers. Google must realize that people's desire for privacy should override the company's desire for personal information. This discussion about privacy is both culturally embedded – Western cultures have legal protections of privacy and value one's privacy over and above the concerns and needs of society as a whole – and financially protected – from one's bank account to one's income and wealth. Unlike other countries, individuals file their own tax returns in the US so that their personal data remain under their private control and are released exclusively by them. Though workplaces deduct certain taxes from paychecks (federal, state, social security, unemployment), it is up to the individual to file a tax return annually and to certify its veracity.

When Internet retailers, such as Amazon, maintain accounts for millions of customers and track their buying habits, they may provide a service of sorts – ensuring you remember that you already bought this or that book and giving you an annual record of your purchases; however, they also provide a self-serving service – pushing you to buy complementary items to those already chosen or reminding you that you like DVDs or CDs of this or that genre. In some cases, these Internet outlets put you on an e-mailing list that is even more invasive than the catalogues they send via snail-mail: they appear on your laptop or

DOI: 10.1057/9781137312402

smartphone every few hours with reminders that are difficult to ignore. They know you own a pet and that you purchase about $300 worth of goodies for that pet, and that this month you have been forgetful if not outright negligent. With seductive appeal on the one hand and a sense of guilt on the other, consumers are nudged to make yet another purchase with a click or two, and their chosen item will be at their doorstep a day or two later free of charge. Who can resist this temptation?

Somewhere in between the monitoring of workers in the workplace and consumers in their homes, there is another form of record-keeping that in the Digital Age is the most powerful tool in the hands of post-capitalists: the credit score, or more recently, the e-score. Credit scores vary from country to country, but regardless of their range are used as a means by which to judge the creditworthiness of individuals, based primarily on their credit history: house mortgage, car loan, business loan, student loan, credit card scheduled payment, personal bankruptcy, bounced checks, and other financial transactions. One's past performance is recorded and points are either deleted or added depending on whether bills were paid on time or not at all, whether there was any payment default or prepayment, as well as one's liquid assets and salary. The collection of vast data from varied sources is quite easily undertaken, especially if there is one identifying number (social security) to ensure that all the data are indeed associated with a particular individual. Though American law forbids the use of a credit score as a variable in one's hiring, one's credit history can be examined. Creditworthiness is calculated in statistical terms that predict the likelihood an individual will pay back whatever new loan is being applied for. As can be expected, numbers never tell the whole story, and there is plenty of room for reporting errors (wrong person or wrong incident). Companies, such as Reputation.com, have been established in order to correct credit score mistakes or monitor one's reputation in general – what comes up when a name is entered into a search engine – and also the particular creditworthiness in credit score sites. If once we worried what file we had in the vaults of the FBI or CIA, now we must worry what anyone can say about us and enter into the information storage of the Internet.

In addition to credit scores and one's composite profile in search engines – how often do you "Google" yourself to see what is being said of you? – there are now e-scores that are playing an increasingly important role in e-commerce, that is, the electronically generated marketplace. As Natasha Singer reports in *The New York Times*, the desire to predict as

DOI: 10.1057/9781137312402

accurately as possible creditworthiness has brought about what she calls "a Google-esque business, one fueled by almost unimaginable amounts of data and powered by complex computer algorithms. The result is a private, digital ranking of American society unlike anything that has come before." In fact, "Every month, eBureau scores about 20 million American adults on behalf of clients like banks, payday lenders and insurers, looking to buy the names of prospective customers. An eBureau spinoff called TruSignal, also located here, scores about 110 million consumers monthly for advertisers seeking select audiences for online ads" (Singer 2012, 18). Unlike other credit scores that compute one's past credit performances, the new algorithm accounts for other variables as well, such as one's occupation and the neighborhood in which one lives, one's previous propensity to spend on luxury goods or travel abroad. According to Singer, "a growing number of companies, including banks, credit and debit card providers, insurers and online educational institutions are using these scores to choose whom to woo on the Web. These scores can determine whether someone is pitched a platinum credit card or a plain one, a full-service cable plan or none at all. They can determine whether a customer is routed promptly to an attentive service agent or relegated to an overflow call center" (Ibid.). So, why worry about this refined evaluation of one's creditworthiness? Isn't the marketplace entitled the freedom to pursue whatever venues to ensure prompt payment and loss prevention?

What is at stake is not simply another instrument for evaluating individual consumers, but the de facto creation of "a new subprime class: people who are bypassed by companies online without even knowing it." It's the secrecy of this process that isn't reporting scores or letting individuals know that they are being assessed to begin with (unlike credit reports about which consumers are notified). The new e-score is "derived from nontraditional data and promoted for marketing." Using "lead-generators," companies, such as eBureau, are able to collect data indirectly from consumers. "Lead generators are companies that set up consumer-friendly Web sites with the goal of funneling potential customers to businesses ranging from financial institutions to wedding photographers... sites like Bankrate.com, for example, offer rate calculators and other tools that prompt people to fill out forms with their names and contact information. The sites then transmit those consumers' information to mortgage brokers, credit card issuers, car insurers and the like, offering access to these prospective customers, or leads, in return for a

DOI: 10.1057/9781137312402

finder's fee." With a few hundred of thousand leads, one can obviously begin to compile credit profiles. This is how the system works: "A client submits a data set containing names of tens of thousands of sales leads it has already bought, along with the names of leads who went on to become customers. EBureau then adds several thousand details – like age, income, occupation, property value, length of residence and retail history – from its databases to each customer profile. From those raw data points, the system extrapolates up to 50,000 additional variables per person. Then it scours all that data for the rare common factors among the existing customer base. The resulting algorithm scores prospective customers based on their resemblance to previous customers" (Ibid.).

The issue is not only that there is no federal oversight and potentially unfair treatment of individual consumers, but the inability to correct mistakes or misinterpretation of the data collected about the individual. It's not simply that the legal system lags behind digital innovations, as Davidow reminded us, but that to some extent the marketplace insulates itself from the social framework within which it operates, and more profoundly, it does so in secret. When consumers know their credit scores and share them with banks and lending institutions, they are collaborating in the diffusion of information about their own financial status and history. With secrecy as the operating principle according to which one's worthiness is evaluated, one is reminded of the horrors of the KGB in the former Soviet Union, the nightmare against which the Austrian economists fought. While they thought the marketplace would be neutral in judging individuals, considering them equally worthy potential consumers, the Digital Age is proving otherwise: individuals can be identified and isolated from the group, quarantined, so to speak, in consumer purgatory. We are not simply aggregated according to how many "friends" we have on Facebook or in terms of our "Prime" status at Amazon; instead, we are classified in terms of our consumer worthiness, if our position in the marketplace warrants attention at all, and if yes, at what rate and under what conditions.

As we think of how this might be appropriate for entrepreneurship – anything that is not illegal is permissible – more services keep popping up that relate to the monetary value of our web-behavior. Though the majority of Americans are leery of advertisers, or for that matter, anyone tracking them, a start-up by the name of Enliken encourages consumers to sell their own data. "Data exhaust" is like "carbon footprint" in the sense that digital traces are left behind or are generated when you

DOI: 10.1057/9781137312402

surf the web. This exhaust has obvious value for companies that would like to know if you are a worthy target for their products rather than waste their time trying to sell you perfume if you have never bought it or never checked out websites that sell it. This idea of selling one's own data empowers the individual with a choice and even some cash along the way rather than being spied on by others who then sell the information to potential retailers. Though the founders of Enliken pay only $1/month to individuals who are signed up, the point they make is that it's free money that otherwise you wouldn't have had even though your data would have been tracked by a third party (Bruder 2012). One could make the same argument about self-exposure (instead of concern with privacy) in the case of Facebook and other social networks that one signs up for and to which one voluntary uploads personal data.

III Entrepreneurship

As I have shown elsewhere (Sassower 2009, ch. 3), venture capitalists emerged in the 1960s and 1970s as major contributors to the development of what we call nowadays start-up companies, focusing primarily on the nascent computer industry. From the Boston area with its concentration of academic institutions to the Silicon Valley in California with another set of academic institutions, venture capital (VC) investments proliferated well into the beginning of the twentieth century. By the end of the twenty-first century a new phenomenon emerged: angel investors (AI). Unlike their predecessors, they were more organically tied to the people and companies they helped. At times they were related to them; at others their attitudes toward the returns they expected from their investments differed from VCs. It's not quite a philanthropic affair, but it comes close to it. Absent a ruthless expectation for success and huge monetary payoffs, AIs nurture the companies they help fund, providing personal expertise and allowing ideas to incubate for as long as it takes (rather than rushing them to the marketplace). Though it seems as if AIs have come to the scene after venture capitalists, one can find multiple forms of patronage dating back to the Renaissance, where wealthy patrons subsidized individuals with great ideas and no funds to bring them to fruition. Another characteristic differentiation is that while some estimates put the average investment by VCs at $7 million, the average AIs is about $30,000. These are rough numbers that just illustrate the magnitude

DOI: 10.1057/9781137312402

associated with either group of investors. Some estimates claim that both sets of investors funded about $20 billion each in 2010 with the difference being that while VCs funded about 1,000 companies, AIs funded more that 60,000 (because of smaller amounts in each case). Whether one set of funding precedes the other depends wholly on individual cases, and cannot be said to necessarily follow a predictable pattern of first VCs and then AIs or vice versa.

There are other entrepreneurs who take advantage of digital technologies in the financial sector. The now common (and about to be regulated) practice of high-frequency trading (not under consideration here) takes advantage of high-speed computers and programs, trading millions of shares in a fraction of a second in ways unknown only a decade ago. These kinds of manipulative exercises may be extremely lucrative for a few hedge-funds or investment banks that can leverage daily billions of dollars, but in fact serve no purpose in expanding our knowledge base or the progress of digital technologies. Instead, they register very short-term arbitrage gains on behalf of large institutions that can afford the software expertise and the enormous investment in the hardware that processes them. These transactions bring entire exchange markets to their knees with processing overloads whose results remain below the radar, so to speak, until a catastrophe happens with so-called rogue traders (The London-based Whale of JP Morgan Chase) whose large bets were off and caused a loss of $2 billion (or as much as $5 billion) in the course of a few hours in 2012. Such cases of high-frequency trading, the speed and extent of which might be shortly curtailed, are of no interest under the rubric of entrepreneurship because they engender no economic growth outside of the immediate profits (or in some cases great losses) to few individuals or hedge-funds. The impact, when felt at all, is in cases of losses where tax policy ensures public participation in the writing off of these losses against past or future gains. It's not that such algorithms have not been introduced to the stock markets since the 1980s, but rather that their processing powers have increased so much so as to move them from a change in degree – faster and faster – to a change in kind – allowing only a handful of extremely well-financed entities to participate in this gambling game that in itself changes the value of stocks, options, and commodities by "making" markets in split seconds that don't add to liquidity but only to penny-size differences. For more details, see the comprehensive analysis by two critics who practice in this area (Arnuk and Saluzzi 2012).

DOI: 10.1057/9781137312402

What's more fascinating is the transformation of the entrepreneur funding model that has been brought about with the advance of digital technologies. At stake here is not the standard investment in digital technologies – whether through government subsidies or VCs or AIs – but a reversal in the very conception of the relationship between the inventor and the investor. Perhaps in broad terms this is embedded in what Shirky (2010) calls "cognitive surplus," which for him can be gained from an ongoing progressive historical march of innovations all the way from the Gutenberg press to the Internet. Under conditions of increased connectivity among people, there is the potential and actual proliferation of ideas and innovations because more access translates into a broader market of ideas, especially when these ideas are provided most of the time free of charge. This optimistic belief in the progress of history that is bound to provide an ever-larger basket of digitized fruits may be the backdrop against which the relationship between investors and inventors is being transformed.

What has become known as crowd-funding (CF) is a process by which individuals from across the globe fund projects from across the globe, with small amounts or large, in any area that might be of interest. There are different forms of CF: some expect something in return, a smartphone application in return for a $20 investment that, if enough people sign up for, will be developed; a CD with music in a specific genre for which thousands have basically prepaid; others look like a VC but feel like philanthropic enterprises. The largest of these websites is Kickstarter, which tells its own story: "Kickstarter is a funding platform for creative projects. Everything from films, games, and music to art, design, and technology. Kickstarter is full of ambitious, innovative, and imaginative projects that are brought to life through the direct support of others. Since our launch on April 28, 2009, over $350 million has been pledged by more than 2.5 million people, funding more than 30,000 creative projects." How does Kickstarter work? "Each project is independently created and crafted by the person behind it. The filmmakers, musicians, artists, and designers you see on Kickstarter have complete control and responsibility over their projects. They spend weeks building their project pages, shooting their videos, and brainstorming what rewards to offer backers. When they're ready, creators launch their project and share it with their community." So, this means that "every project creator sets their project's funding goal and deadline. If people like the project, they can pledge money to make it happen. If the project succeeds in reaching

DOI: 10.1057/9781137312402

its funding goal, all backers' credit cards are charged when time expires. If the project falls short, no one is charged. Funding on Kickstarter is all-or-nothing."

Why is funding "all-or-nothing"? According to the website: "All-or-nothing funding is a core part of Kickstarter and it has a number of advantages: *It's less risk for everyone.* If you need $5,000, it's tough having $1,000 and a bunch of people expecting you to complete a $5,000 project. *It motivates.* If people want to see a project come to life, they're going to spread the word. *It works.* Of the projects that have reached 20% of their funding goal, 82% were successfully funded. Of the projects that have reached 60% of their funding goal, 98% were successfully funded. Projects either make their goal or find little support. There's little in-between. To date, an incredible 44% of projects have reached their funding goals" (emphasis in original).

The interesting part of this new model for entrepreneurship is the fact that creators have full control of their projects, unlike VCs and AIs that interfere, redirect, and even take over projects they fund when they are dissatisfied with the speed or direction a project is taking. The point is for backers to support projects so as to help them come to life, regardless of profits. Kickstarter itself receives fees of 5 percent of the total funded project only if the project comes to fruition; otherwise it receives no fees. The rationale for accountability is that the very process of funding a project is public instead of the secrecy associated with Non-Disclosure documentation of VC funding; individual backers have ongoing direct access to the project creator. Like any other speculative investment, there is neither guarantee of success nor a way to refund failed projects. Perhaps what distinguishes CF is a basic trust between creators and backers, using a single platform to convince each other that they are all working in good faith to accomplish what they couldn't have otherwise, as creative individuals and as investors.

Perhaps such websites will turn out to be the standard of entrepreneurship in the Digital Age. CF's $350 million pales in comparison to the $40 billion combined investments of VCs and AIs, but it's an idea that is working itself into the curious minds of creators and backers alike, an opportunity to know what is happening around the world and a way to participate in it with minimal sums of money. Certain barriers to entry, as the economic term suggests, have been lowered if not eliminated. This alone will not make the next Microsoft come to life, but perhaps it will level the playing field of creative minds, allowing more

DOI: 10.1057/9781137312402

of them to enjoy the potential success of a postcapitalist marketplace in the age of postmodernity, where digital technologies challenge not only material conditions but also conceptual frameworks. *Along the way, CF also challenges standard models of capitalism because it isn't exclusively focused on profits or standard models of socialism where government subsidies fund collective welfare. CF exemplifies the postcapitalist ethos that celebrates collaboration and planning with the initiative of individuals to excel and succeed in accomplishing their dream project.* Will the creators remain grateful to their backers? Will they feel any obligation to reward their backers when becoming hugely successful? Since the CF experiment is so new and about to be more fully regulated, it's too early to attempt answering these questions.

DOI: 10.1057/9781137312402

4

Legality and Morality: Intellectual Property, Virtual Currency, and Corporate Responsibility

Abstract: *To better appreciate the moral hazards associated with digital technologies, it's useful to examine the debates over intellectual property rights, their legal and political frameworks as applied to a postcapitalist marketplace. Another litmus test is the introduction of virtual currency, "bitcoin," and its potential to overcome banking controls. Finally, corporate responsibility will be discussed in light of recent financial developments.*

Sassower, Raphael. *Digital Exposure: Postmodern Postcapitalism*. Basingstoke: Palgrave Macmillan, 2013. DOI: 10.1057/9781137312402.

I Intellectual property: rights/duties

As seen from the perspective of crowd-funding, intellectual property seems completely unproblematic: the creator or inventor maintains full control over the invention and can have it registered for copyright protection. What becomes more interesting in challenging the classical paradigm of capitalist control over one's invention (as private property) is when there is no attempt to control one's invention, as in open-source cases where whatever is invented is offered, free of charge, to anyone. This is, of course, in stark contrast to the current state of affairs where billions of dollars are spent on litigation rather than on research and development both to protect one's intellectual rights and to thwart any attempt to compete, fair and square. This discussion can remain within the domain of legality where lawyers contest any attempt to infringe on one's copyright or patent (Lessig 2001, 2004); or can be broadened to incorporate ideas about our political framework (freedom, community) instead of focusing on the proliferation of a capitalist mind-set into the realm of ideas rather than things (Berry 2011). Historical surveys that trace American concerns with patents and copyrights as framed by the US Constitution and misapplied in terms of British ideas about the Commons are relevant here as well (Benkler 2006, Sassower 2009).

Lawrence Lessig, a well-known legal scholar, writes about the new realities associated with the Digital Age and the ways in which we ought to readjust our thinking first, and then our laws, to address these changes. He expresses the need to protect and encourage what he calls "amateur" creativity that inspires young people to participate in what he calls the "Read/Write" culture rather than be limited to the "Read/Only" culture that merely receives but doesn't contribute to cultural creations. This can be accomplished by limiting the reach of copyright law to more "professional" spheres and loosening its reach when it gets to the free-wheeling "re-mixing" artists enjoy under the legal concept of "fair use" (Lessig 2008, 33). His ideas regarding the limits that should be placed on the ability to copyright everything is phrased forcefully in these terms: "if copyright regulates copies, and copying is as common as breathing, then a law that triggers federal regulation on copying is a law that regulates too far" (Ibid., 269). And this overregulation has become more noticeable in the age of digital technologies. Economically speaking, since the bottle-neck that can be controlled is that of distribution (records is the example being used), once new technologies subvert old analog models,

DOI: 10.1057/9781137312402

the whole "content industry" is threatened (Ibid., 38–40). Digital technologies provide broader and cheaper venues of free speech expression than ever before, and the law has to adjust to this change (Ibid., 83). The legal claim here is that "we need to decriminalize creativity before we further criminalize a generation of our kids" (Ibid., 114).

In case Lessig's proposals sound outlandish and even counterintuitive in a culture that fosters private property rights, he seems to highlight the kind of practices one would expect to find in a postcapitalist marketplace where collective sites and archives are publicly owned and therefore free of charge, while other individuals pursue fame and profit the old-fashioned way, protecting their inventions from open use by others. Lessig counts all kinds of examples of the "software sharing economy," from Project Gutenberg, Distributed Proofreaders, Distributed-computing projects, The Internet Archive, The Mars Mapping Project, The Open Directory Project, to Open Source Food as a series of websites and digital archives open for the free use of consumers (Ibid., ch. 6). For him, the hybrid economy combines the commercial economy with the sharing economy, using free software as the paradigm. He distinguishes between "community spaces" which are virtual spaces for sharing information, "collaboration spaces" where a specific task, goal, or work is undertaken by many people, and the third type of "communities," such as Second Life, where one chooses to "live" among others. This illustrates the varied ways in which the Digital Age expands on the pre-figured confines of capitalist markets where buyers and sellers confront each other under different conditions and with different ends in mind (Ibid., ch. 7). This may give an impression that the gray areas that postcapitalism endorses will eventually blur the distinction between for-profit commercial enterprises and public-works ones where free sharing is the standard. But this would be a mistake, according to Lessig. Though the values are different between sharing and commercial economies, it's important to keep them apart, otherwise what seems like a "gift economy" will fall back to one of profit, and will undermine the "culture of generosity," turning it into that of profit (Ibid., 232–233). In other words, if the two "economic cultures" coexist, they retain their respective uniqueness and remind any observer of their differences.

Though writing in the same time about the same issues concerning intellectual property rights and duties, David M. Berry goes much further than Lessig in his assessment and proposals concerning the political dimension associated with what seem to be capitalist economic interests

DOI: 10.1057/9781137312402

enshrined in legal codes and regulations. His is an attempt to "uncover the way in which the open source and free software groups are challenging our existing liberal categories (around cultural production, knowledge ownership and authorship) both in economic terms (that is, as a new form of commons-based peer production) and in terms of political liberties (for example, the question of free speech in a democracy)" (Berry 2011, x). Unlike Lessig, Berry follows a different path of understanding political economy holistically and historically, appreciating government intervention when needed, and with moral considerations that date back to eighteenth century thinking. He does agree with Lessig that the gift economy where standard capitalist variables are suspended has become more noticeable publicly rather than limited to exchanges among family members and friends (Ibid., 7–9).

Berry follows Benkler and Lessig in asking for a core commons infrastructure that is not simply technically available and protected from market intrusion, but is also politically safeguarded as a value in itself (Ibid., 21–24). In doing so, he is part of what has been recently termed "digital environmentalism," a movement that draws its inspiration from the environmental movement both in legal quarters (suing to end pollution) and in its widespread acceptance by the public (from recycling efforts to the preservation of land). If digital environmentalism adheres to some of the principles and strategies already in place by environmentalists and ecologists, there is reason to believe that the classical notion of a commons will become more broadly understood by a public used to obeying the strictures of intellectual property constrains, as some suggest (Boyle 1996).

Berry is worried about a "global intellectual property regime" (2011, 31) insofar that on the one hand, national legal restrictions are being enforced globally, and on the other, societies whose notions of intellectual property differ from our own will readily adopt our standards. This is the kind of global capitalism with legal teeth of which he is fearful. He argues that "the ownership of copyrights and other intellectual property rights will be a site of power *and* resistance in the twenty-first century" (Ibid., 32, emphasis in original). Power consolidation is indeed a characteristic of global capitalism or "Supercapitalism," as Robert Reich terms it (2007), one that becomes so powerful and removed from individual decision-makers and the public that any form of "resistance" envisioned by Berry remains wishful thinking (unless exercised in small-scale personal action, such as refusing to connect to the Internet).

DOI: 10.1057/9781137312402

Berry is correct in rehearsing the classical view of copyright extension beyond the original fourteen years ascribed in the Constitution, suggesting that "corporations increasingly seek to own and profit from creative copyrights in perpetuity, in effect owning and controlling culture... By controlling the use of trademarked phrases, brands or copyright claims, code is able to act as a gate-keeper to knowledge and police the ability to use and disseminate it – through code-enforced intellectual property rights (such as digital rights management). Computer code has the potential to have a chilling effect on democratic debate" (Ibid., 37).

For Berry the question isn't economic in nature even though it finds its most extreme expression in the corporate nexus of profit-optimization and power consolidation; instead, it immediately becomes a political question of democracy and the institutions we develop to ensure that its principles will not disappear in the marketplace. Following Steve Fuller, he wants to have a politics of the commons such that political activity in Hannah Arendt's sense (where the public domain is the arena for public debate) could flourish, as resistance to the oppressive and profit-hungry political economical system that devours every facet of code for its property rights (Ibid., 198–200). This, of course, has global implications, as we have seen at the end of the previous chapter.

In an age where Swiss cows are implanted with chips that would alert the farmer of their menstrual cycles, where Stanford University's pathologists design molecular imaging techniques to penetrate the human body and expedite tests, and where patients can alert their doctors to their mood swings or diabetic patterns, the question of intellectual property rights shifts from the entertainment and media domains to those of personal and public health (see also Topol 2012). Under what conditions should there be any constraint to access any of these life-saving techniques and devices? Aren't they fundamentally public treasures whose expression happens to be in the hands of this rather than that "inventor"? In other words, if it can be argued that all knowledge claims are indeed collectively amassed and produced, requiring the contributions of generations of thinkers and doers (Sassower 2009, chs. 2–3), then it seems quite arbitrary if not silly to have artificial cut-off points where one's ideas and thoughts, insights and experiments, suddenly become one's own. Does anyone really own an idea? Does anyone really own a technique? Or are they inherently public insofar as they have benefited from the wisdom of others, from the basic education one receives to the particular institutional context in which one works? With these

DOI: 10.1057/9781137312402

questions in mind we can shift to a more radical attempt to eschew the banking industry in the name of providing services that are ultimately controlled by end-users rather than a central authority.

II Virtual currency

Just as new digital technologies have challenged the classical understanding both of the (originally British) Commons as a public-access space in modern society and intellectual property rights as enshrined in the US Constitution, so has the emergence of virtual currency challenged the dominance of central banking in democratic nation-states and the classical interpretation of the role money plays in capitalist society. Virtual currencies have been around for a while in different manifestations, from those used in games, like Monopoly, to more sophisticated video games where "money" is earned when one moves from one stage to another; likewise, they are used in closed-communities where exchanges of coins or tokens are prevalent. None of these challenges the very foundations of the capitalist enterprise and its institutional protection, legally and morally. E-money or virtual currency tests the limits of the capitalist-state system. And here is where the postmodern perspective might be of help.

According to Lyotard, the way in which knowledge claims are legitimated within modern institutions is through "language games," discourses that include certain utterances according to a set of rules that are themselves self-legitimating (1984, 9–17). *It's the goal of these institutions, such as universities, state agencies, banks, the military, to retain control over the language games played within their purview so as to minimize dissent or communication breakdowns.* While drawing the boundaries of their respective discourses, institutions control the kind of inventiveness that bespeaks of progress – new ideas, novel experiments, and experimental policies, all of which would be considered "moves" within the confines of the language games. Perhaps the games themselves (institutionally sanctioned) can never be *replaced* – completely done away with; yet they may be open to ways of being *displaced* – providing (institutionally sanctioned) alternative modes of communication and interaction. In this context, this process can work in a democracy with institutionalized guarantees for maximal disagreement which should be reasonable and civil. In short, there has to be built-in political latitude and legal support protecting the process.

DOI: 10.1057/9781137312402

This postmodern way of thinking about knowledge in general and of institutions as repositories of knowledge in particular helps set the stage for what has been attempted since the 1990s in the area of virtual money or e-Money. The question that haunts postmodern postcapitalists remains: is this novelty merely a "move" in the financial game? Or, as some advocates of virtual currency claim, is this a novel "game-changer" that deserves a fundamental rethinking of our political economy? Whether called "digital currency," "e-Money," "electronic cash system," or any other term, digital technologies have pushed the conceptual and practical limits of what we commonly understand as "money." As the classical account of money suggests, it has both a value and a function, being exchanged for actual commodities or services according to a certain cultural logic that substantiates its ultimate legitimacy (and desirability). As Georg Simmel suggests: "As a tangible item money is the most ephemeral thing in the external-practical world; yet in its content it is the most stable, since it stands as the point of indifference and balance between all other phenomena in the world ... Money expresses the relationship that exists between economic goods" (1978/1900, 511).

Time and again, Simmel reminds his readers that money as such isn't reducible to the particular coins or bank notes with which they are familiar. He emphasizes the ephemeral character of money and its repository of the expression of human relationships with each other and with objects in their economic and cultural world. This leaves an opening, more than one hundred years later, for the development of digital currencies, ephemeral as they are, yet suggestive of the relationships individuals have developed in virtual exchanges. Even more poignant in the current discussion are Simmel's parting words at the end of his definitive treatise on the history of money: "Money, as an institution of the historical world, symbolizes the behavior of objects and establishes a special relationship between itself and them. The more the life of society becomes dominated by monetary relationships, the more the relativistic character of existence finds its expression in conscious life, since money is nothing other than a special form of the embodied relativity of economic goods that signifies their value" (Ibid., 512).

The point of departure for Simmel and for contemporary cyber-experts is where culture is and what relationships humans have with each other: are we meeting at the public square and expect a commitment to a Commons of sorts or rather in virtual space that allows for anonymity and disregard of state institutions? In fact, Simmel's conclusion

DOI: 10.1057/9781137312402

about the historically changing value and function of money within state and commercial institutions opens a critical "door" for those who developed digital currencies. Originally associated with "cypherpunks," a group of cryptographers that counted among its members Bram Cohen who invented BitTorrent, and Julian Assange who started WikiLeaks (Grinberg 2012b, 24), this loosely connected group was intent on challenging government authority in Berry's sense of resistance. In 1998, Wei Dai proposed "b-money" as a digital currency that couldn't be traced or regulated by the government. Ten years later, Satoshi Nakamoto (a pseudo-name) released his version of "bitcoin" (Ibid.). In his manifesto, this anonymous inventor argues that "what is needed is an electronic payment system based on cryptographic proof instead of trust, allowing any two willing parties to transact directly with each other without the need for a trusted third party" (Nakamoto 2009, 1). These early attempts at subversion could be dismissed as marginal or at best a nuisance to the monetary system. Yet, their persistence and the philosophical issues they raise are worthy of some detailed commentary.

A coin is defined by Nakamoto as "a chain of digital signatures" or a "hash." A cryptographic hash function is an algorithm that configures an arbitrary block or string of data and affixes to it a hash value. With a "timestamp" one knows the time-line of each hash so that its movement from one user to another cannot be reversed, thus avoiding double-counting of one's hash or bitcoin. Though users can remain anonymous during the exchange, the exchange itself is public, protecting the privacy of users while keeping them publicly accountable. Those who own bitcoins are required to "mine" them, which means finding a solution to a difficult problem that only high-powered computers and sophisticated programmers can perform. In some cases, bands of cryptographers can collectively mine bitcoins and divide them among themselves (Ibid., 1–9). According to Reuben Grinberg, who provides the best summary of this process and product, by October 2011, "a bitcoin (currency ticker BTC) is worth about two US Dollars (USD), there are about $20 million worth of bitcoins in existence, there are probably around 20,000 Bitcoin users, and over $300,000 worth of bitcoins are traded every day" (2012a, 160–161). The bitcoin ecosystem includes "exchanges, transaction services providers, market information and chart providers, escrow providers, joint mining operations and so on" (Ibid., 165). Surely this ecosystem is comprised of earnest cryptographers who wish to show off their expertise and petty criminals who are looking for ways to hide from scrutiny,

DOI: 10.1057/9781137312402

speculators who believe that there is a market in everything exchanged and those worried about the loss of currency values since the abolition of the gold standard under president Nixon in 1971, as well as those considering the Internet a source of ongoing cerebral entertainment.

There are multiple issues addressed by advocates of digital currencies: first, to have a "peer-to-peer electronic cash system" (Ibid.); second, to eschew "third party" reliance, primarily government central bank or any other commercial bank; third, to replace "trust" with cryptography; fourth, to avoid the collapse of money value because of inflation (ongoing money printing); and fifth, to avoid transaction costs levied by credit card services, banks, or other Internet providers, such as PayPal. *These five issues can be rewritten as: first, individual freedom; second, distrust of government; third, trust in technology as a neutral and fair method; fourth, distrust of capitalism; and fifth, distaste for waste.* An overarching ideology that engulfs these concerns is libertarianism or anarchism, both of which prefer reliance on individuals and the minimal role of government or any of its agencies in human affairs. The motives may have been at first more closely tied with the desire to show off personal expertise and ingenuity in cryptography, defying government control over a set of technologies that ought to be freely used rather than restricted for military or surveillance operations. Later they evolved into a more thorough dislike and distrust of government agencies in general, but instead of standard hacking practices that attempt to disable government websites or computers, it is an attempt to provide a parallel solution to transaction problems. And finally, with exchange sites, such as Mt. Gox, there is a genuine goal of providing alternative modes of money exchange, displacing the hegemony of government-issued notes and securities without ever achieving a full replacement.

As a *process of displacement* – offering ways to trade electronically with bitcoins – this novel digital currency exemplifies a postmodern mindset: using novel techniques, it challenges the status-quo. But is it in fact a game-changer? Though different from other virtual currencies that are found in the virtual world of games, such as Farmville and Second Life, where transactions in Linden Dollars has reached $50 million on a monthly basis, bitcoin wants to appeal to merchants who trade through the Internet but who are selling actual consumable goods. The Linden Dollar is in principle no different from Monopoly money that is useful only within the confines of the game, unless your family or friends are willing to let you buy a cookie or beer with them. Virtual gaming opens

DOI: 10.1057/9781137312402

participation horizons unimagined by Monopoly players sitting on the floor and rolling the dice, but in principle the rules of the game as they pertain to the value of money as a means of exchange are internally fixed, and hardly affect external markets. Of course your local merchant could hypothetically sell you a pair of shoes for Linden Dollars or Monopoly bills, but this an unlikely event.

More profoundly, then, we are at a point to ask: Does bitcoin and any other digital currency really undermine the authority of the federal mint to print money? With claims to safety and reliability, trustworthiness, and increased use, the bitcoin experiment is limited to about $20–40 million and will never threaten the status-quo of trillions of dollars in annual transactions. This is not simply because there are various legal concerns associated with its proliferation (Is it counterfeit money? Is it a way to launder money? Is it taxable?), but because of its relative insignificance in the economy. *Perhaps the valuable lesson of the bitcoin experiment is the realization that even the postcapitalist framework depends on legal rules for its stability – rules of the game that dictate certain moves and expressions, certain boundary conditions.* While national currencies, and more problematically the Euro, rely on government policies that protect and bolster (or deflate) their value when economic conditions warrant such activity, all of these decentralized digital currencies are susceptible to fraud and attack, loss of value, and charges of forgery and imprisonment. This is not to suggest that currencies cannot be floated without government backing or the backing of gold, for example, as is evidenced with the introduction of the so-called Swiss Dinar in Iraq. Though backed by the Iraqi government before the 1990 Gulf War, it was “disendorsed” after the war and a “Saddam Dinar” was introduced in its place. The fascinating fact is that in northern Iraq the Swiss Dinar retained its trading value while the Saddam Dinar was affected by hyperinflation after the 2003 invasion of Iraq (Grinberg 2012a, 174–175). Nor is it to suggest that legal reasons alone warrant the closing of the potential for alternative currencies in the Digital Age. Instead, what becomes clear is that digital currencies, such as bitcoin, are different only *in degree* from other currencies, and do not yet provide the difference *in kind* that changes the economic game of financial transactions and institutions. This is lamentable because the Digital Age could potentially alter our view of and our activities within the world not only in postmodern ways (providing more alternatives), but in substantive changes of power relations.

DOI: 10.1057/9781137312402

III Corporate responsibility

There are various ethical issues that arise when surveying the pervasive use of digital devices, from justice and fairness to privacy and theft. Perhaps we should also recall the ancient Greek insistence on *prudence* as a way to approach these questions, perhaps more reasonable in economic terms. From an economic perspective, are our moral standards to be tested in the Digital Age differently from earlier ages? For example, the straightforward condemnation of theft in material terms (stealing an apple), however circumscribed under special conditions (to save the life of a child), is arguably more complex in an age of ethereal theft (on the Internet). Though the principle remains the same ("thou shall not steal"), whether understood in the biblical sense (a divine injunction in the Ten Commandments) or in the Kantian sense (universalizability from one's case to all others), it is paradoxically less gray in the Digital Age than expected. Under moral framing, Internet theft is no different from all other forms of theft. This suggests that two interrelated topics should be examined: first, how the fight for intellectual property rights of large corporations, such as Apple, has transcended legal concerns into the domain of economic power-play, and second, how corporate might has procured digital technologies to gain advantage over all other participants in the marketplace, such as the use of high-frequency trading in all stock exchanges. Though different in many ways, these kinds of practices give pause to any idealized hope in the ability of postcapitalism to find a hybrid middle-ground between capitalism and socialism.

In August 2012, Apple was awarded a $1 billion patent infringement judgment against Samsung by a California court. Supposedly a result of a landmark decision, this award to protect the exclusive sale of Apple's iPhones hasn't decreased the sale of Samsung's competitive Galaxy Smartphone. In October 2012, a Dutch court found the opposite to be true, namely, that Samsung hasn't infringed on Apple's patent in regards to its touchscreen technology. And in November 2012, Apple and HTC agreed to collaborate and settle their respective lawsuits. Will lawsuits ever cease? When Apple's net income is over $8 billion in one quarter in 2012, while its annual revenue exceeds $156 billion, should we concern ourselves with its litigation practices? What these cases have done, though, is draw more attention to the age-old argument about intellectual property rights, including patents and copyrights (as opposed to trademarks which broadly deal with branding and not specific technologies or inventions). Standard arguments pitted those advocating for copyright protection (on the economic grounds of incentives to inventors and corporate research and development expenditures) against those arguing for abolishing such protections (because market forces should permit free usage of knowledge and ideas as a public good for which society already paid). The new twist in this debate is more focused on legal costs. More specifically, some estimate that Apple and Google have spent in 2011–2012 about $20 billion on litigation and patent application, more than they have spent on research and development of new technologies or devices. No matter how much pressure is put on the legal system to reform the patent process or revisit the legal framework for patent protection – how many years? what scope? how broad the proposal? – as long as large corporations have the financial wherewithal to fight against small startups, the idea of a free market open to all newcomers remains an ideal if not a sham (Duhigg and Lohr 2012). Instead, *this form of postcapitalism enjoys the political and legal protection of well-established businesses so that their well-being is guaranteed while disregarding the marketplace as a forum for competition and innovation.* Corporate powerhouses protect themselves from market pressures rather than dominate the market in light of challenges from newcomers.

Monopoly formations under classical capitalism usually enjoyed economies of scale in production and distribution. That is, because of their size, large companies could be more efficient and lower their marginal costs substantially – large assembly lines, comprehensive and multilayered distribution centers with sophisticated technologies (the legend of the original Wal-Mart logistical advantage). They became larger organically: they grew to become monopolies by fairly competing with others. According to this model, there are no barriers to entry into the marketplace, and everyone gets a fair shot at competing with everyone else. What has changed in this postcapitalist market is the realization that what we are dealing with is a political economy; namely, an economic framework set in place and protected by a legal framework, one that through lobbying efforts can minimize competitive exposure. Perhaps there is justification for monopoly formation and dominance, but it is coming from legal experts rather than from economic realities. Is Apple keeping its market share and profitability because of its legal activism or its superior technology and design? Though legally defensible, this kind of corporate behavior seems unethical: it undermines the freedom to compete, fair and square, and

DOI: 10.1057/9781137312402

revokes the ideal of equal opportunity so dearly espoused by politicians and jurists.

The second way in which corporate behavior in the postcapitalist marketplace has come under attack and deserves moral scrutiny is in the financial industry. Not only the egregious use of high-frequency trading undermine faith in the stock exchanges, as mentioned in Chapter 3, but all other unfair advantages deployed by financial institutions and hedge-funds may pass legal tests. Such legal tests overlook the more long-lasting and profound damage this behavior has on the faith Americans have in financial markets. If investment bankers and hedge-fund managers can play by their own rules with impunity, why should anyone else behave ethically? If personal enrichment is the sole purpose of setting up publicly traded markets for mortgages and currencies, for example, do they deserve legal protection similar to that afforded the neediest of our society? If greed runs Wall Street, should Main Street support it? This is a zero-sum game (despite arguments to the contrary by those promoting infinite growth), where one's gains are someone else's losses, where large losses are so large that government intervention is needed. This makes sense when viewing our political economy as a hybrid of political ideals and economic realities; but it makes less sense when there is no political price to pay for economic gains, or when economic gains remain structurally skewed in the direction of small financial elites.

In recent years, complaints about Wall Street were lumped into three main categories: undue political influence (the revolving door of Goldman Sachs chiefs as political leaders in the Department of the Treasury), federal bailout after the collapse of the mortgage bubble, and compensation to corporate CEOs that is larger than the federal taxes these corporations pay. What is missed when these three areas are scrutinized (and of course they deserve public scrutiny and moral outrage) is a more routine misconduct that has become part of a new postcapitalist reality: "pay as you go" legal fees. As Michael Rothfeld reports (2012), when it comes to corporate crime, we have a different standard by which to judge or mete punishment: "Known as 'deferred prosecution' or 'nonprosecution agreements,' these deals let the U.S. negotiate fines, put companies on probation and get them to change practices." With 1 percent of the population in prison, with overcrowding and shrinking state and federal budgets, one must be sympathetic to any judgment that excludes prison time. The standard questions of deterrence versus fairness may fall by the wayside if we can neither afford to staff prisons nor have enough cells for inmates.

DOI: 10.1057/9781137312402

What also becomes clear under such circumstances is the important role government agencies must play to ensure the smooth operation, even security, of the marketplace. It was Bertrand Russell who after World War II wrote about what functions the government ought to play: security, justice, and conservation (1949, chs. 2 and 5). Economic security was understood by him to be just as important as leaders today speak of national security (foreign relations and domestic policing), and can only be undertaken by the government rather than left to the whims of the marketplace. Indeed, government intervention with rules and regulations, incentives and tax breaks, is crucial to maintain the integrity of the marketplace. *Can the current mind-set of legal authorities when confronting corporate crimes on Wall Street be considered "too big to indict," reminiscent of the justification for bailout under the rubric "too big to fail"?* Barclays (headquartered in the UK) had no problem paying $453 million in fines to American authorities after a criminal probe to its Libor fraud in 2012. Are the bank's activities in the US so profitable that such a fine is just another "cost of doing business"?

Barclays is in good company, though. JP Morgan Chase, Deutsche Bank, UBS, and Credit Suisse Group have reached similar "agreements" in the past few years. They seem to have become acceptable forms of government enforcement of regulations that are simply ignored. Apparently, self-policing isn't even on corporate agendas, and fines are no deterrent punishment, since rules keep on being violated by other financial institutions. Compliance isn't even expected, given the frequency of violation, and the casual response by the offending institutions. Gibson Dunn & Crutcher, a law firm that follows these agreements, claims that the Justice Department has collected over $31.6 billion in fines since 2000 from some 230 agreements (Sassower 2012). What do these numbers mean? How should we interpret them? Obviously, there are ways to use financial "might" to ward off legal "right," paying fines and promising to reform without actually admitting to any wrongdoing. It was Rousseau who argued against the Enlightenment thinkers who believed in the unfettered progress of civilization: the study of the arts and sciences makes us clever but not moral. He would agree that Ivy-League graduates who run Wall Street can argue their way out of any moral quandary without realizing that they have lost their humanity, the compassion for other people and their needs, and of course their moral compass (Rousseau 1964/1750).

DOI: 10.1057/9781137312402

These kinds of routine moral failings bring to mind a corporate culture that disdains the strictures of the rest of society. Most specifically, the enormous wealth and power concentrated in financial institutions on Wall Street (and their global counterparts) insulate them from the legal and moral pressures the rest of society must adhere to. *Postcapitalist or not, postmodern or not, digital or not, corporate behavior should not be exempted from some of the basic tenets that form the Social Contract.* We all live in and expect the same institutional protections we deem necessary for survival and prosperity. We cannot pick and choose what government rules we like – patent protection – and dislike – constraints on monopoly pricing. Outright fraud is relatively easy to detect in the Digital Age even if a bit harder than perpetrating it; but this is on the margins of corporate activity. It's the routine disregard for the rule of law that is more pernicious and therefore deplorable. *When financial institutions routinely pay fines so as to continue their illegal behavior, there is indeed a change in kind and not only in degree of immorality. And here, digital culture is neither a necessary nor a sufficient condition for successful evasion; it simply lubricates the wheels of evil.*

DOI: 10.1057/9781137312402

Conclusion

Abstract: *While technophiles keep on promoting the importance of pervasive success of digital technologies, technophobes maintain that, first, digital technologies are hazardous, and second, that they are hardly game-changers. A critical and balanced approach should make use of postmodern attitudes that are open-minded as well as those of postcapitalist that draw the best capitalism and socialism have to offer. This means, for example, a more nuanced legal apparatus for intellectual property rights that allows large areas of shared information, code, and creative ideas.*

Sassower, Raphael. *Digital Exposure: Postmodern Postcapitalism*. Basingstoke: Palgrave Macmillan, 2013.
DOI: 10.1057/9781137312402.

 DOI: 10.1057/9781137312402

When you count cards in one of the casinos in Las Vegas, the American gambling capital, you'll be asked to leave and barred from any future gambling. On one level, counting cards at Black Jack is cheating; on another, it's just leveling the playing field with the casino that systematically and with government licensing and oversight beats all gamblers as a whole. What is the proper framing of such an impasse between gamblers and the casino? Should casinos be licensed to beat gamblers no matter how good the gamblers are? The so-called house advantage, depending on the game and the skill of the player, ranges from 1 percent to 27 percent. So, why would gamblers keep going to casinos, knowing as they do that the odds are stacked against them? The answer is twofold: some believe that since we are always talking about averages, they can be the outlier winners, while others willingly pay for the entertainment.

It's one thing to talk about a casino, a privately owned entity regulated by a government agency that ensures the odds conform to some "fair" standard, and quite another to talk about the marketplace. Markets are not casinos insofar as they aren't privately owned; they provide the platform for privately owned companies to compete fairly with each other. We'd call the marketplace rigged if we thought some competitors had an unfair advantage before they even entered the market. Assuming our concern remains with the marketplace – of goods and services, ideas and inventions – and not with this or that corporation (or casino), what would ensure fairness in a postcapitalist market? Does the Digital Age provide conditions for fairness and prosperity? Do digital technologies necessarily enhance the growth of the marketplace?

On the one end of the spectrum are the digital technophiles, the ones who truly believe in the great promises of the Digital Age. For example, Peter Diamandis and Steven Kotler, in their book *Abundance,* promote "One Planet Living" as a "global initiative" meant to combat "shortages." Since "humanity uses 30 percent more of our planet's resources than we can replace," and since we want the "lifestyle of the average European, we'd need three planets' worth of resources," or for living "like an average North American, then we'd need five planets to pull it off" (Diamandis and Kotler 2012, 5). Contrast that with: "When seen through the lens of technology, few resources are truly scarce; they're mainly inaccessible" (Ibid., 6). We quickly move globally from scarcity to abundance, and this move is predicated on the extensive use of digital technologies that embody and bring to fruition untold tales of success and prosperity. This eventually translates for them to: "abundance is about creating a world

DOI: 10.1057/9781137312402

of possibilities: a world where everyone's days are spent dreaming and doing, not scrapping and scraping" (Ibid., 13).

This is reminiscent of Marx's utopian vision of communism in the *German Ideology*: "Each can become accomplished in any branch he wishes, society regulates the general production and thus makes it possible for me to do one thing today and another tomorrow, to hunt in the morning, fish in the afternoon, rear cattle in the evening, criticize after dinner, just as I have a mind, without ever becoming hunter, fisherman, herdsman or critic" (Marx and Engels 1970/1932, 53). What strange company they keep without even knowing it, envisioning a utopia where we can all freely engage several marketplaces and freely enjoy the fruits of our engagement.

The exuberance associated with contemporary technophiles finds other promoters in Don Tapscott and Anthony Williams, who urge us to "use the opportunity that the digital revolution presents to rethink and rebuild all of the old approaches and institutions that are failing" (2010, 21). For them, this "opportunity" is global, "one where people with drive, passion, and expertise take advantage of new Web-based tools to get more involved in making the world more prosperous, just, and sustainable" (Ibid., 7). With their terms "wikinomics" and "macrowikinomics" – both heavily indebted to the unexpected success of the open-source Wikipedia – Tapscott and Williams envision "collaborative communities" that transcend boundaries, both disciplinary silos and those of space and time through the Internet (Ibid., 19). Just as Diamandis and Kotler mirror Marx, so do Tapscott and Williams reflect the insights of sociologists of science. When they outline the principles and ideals according to which to organize the "Age of Network Intelligence," they end up repeating Robert Merton's summary of the principles and ideals of the scientific enterprise. Theirs call for collaboration, openness, sharing, integrity, and interdependence (Ibid., ch. 2); his is classified as communism, universalism, disinterestedness, and organized skepticism (Merton 1973, 268–278). To speak of communism is to speak of sharing; to speak of universalism is to speak of openness and interdependence; to speak of disinterestedness is to speak of integrity; and the notion of organized skepticism can be undertaken with collaboration and interdependence. Some two generations apart they all recognize that in order to have a community of researchers and innovators, entrepreneurs and academics, one must adhere to the kind of selfless principles that enhance collaboration across the globe with, as it turned out, the extensive use of the Internet.

DOI: 10.1057/9781137312402

The collaborative elements mentioned earlier by Wilson in Chapter 1 about multilevel evolutionary progress that includes group collaboration are echoed by Barrett in Chapter 2 in regards to international cooperation, and eventually find adherents among the digital technophiles. For them, there should be no barriers to a completely open-source culture where intellectual property becomes public rather than private. What about all the counterarguments regarding the necessary protection of patents and copyright for one's creativity to ensure innovation? Kal Raustiala and Christopher Sprigman, in their engaging *The Knockoff Economy* argue otherwise. According to them, the "monopoly view of innovation" suggests that creators need a monopoly on who can copy their work so as to induce them to create new ideas and bring them to the marketplace for the rest of us to enjoy (2012, 6). Contrary to all the arguments that find economic or financial incentives in relation to one's creativity and its rewards, they coin the "piracy paradox" that shows that "copying actually spurs innovation" (Ibid., 7), especially when it comes to the fashion industry, restaurants and chefs' recipes, football strategy, stand-up comedy, and open-source programming.

This means that instead of having the heavy hand of the law guide innovation paths, a more rapid process of innovation and copying eventually spurs even more innovation because the original, now copied, dies a relatively quick death, and in its stead a new rebirth necessarily ensues. This is not to say that some social norms aren't in place to mitigate inappropriate copying, such as bad taste or poor etiquette, leading the marketplace to shun this copying behavior (Ibid., 78–83). What is significant about this approach is the shift from the *legal framework* that justifies intellectual property claims and stifles the dissemination of new innovations to the *economic arena* where greater exposure to new ideas invigorates the marketplace. This shift illustrates the postmodern mind-set that encourages a multitude of approaches, spurred by digital technologies, without any expectation that there is a preset hierarchy of products and the ideas that brought them to market. *If the vestiges of competitive capitalism are still with us in the postcapitalist era, then the so-called knockoff economy should be thriving. Any curtailing of its potential is too socialist in the bad sense of control and state-planning, where a central authority gives licenses to this creator rather than to the other, chocking creativity as such. To safeguard marketplaces, wider range of options should be granted to any newcomer from anywhere in the world, forcing localized insulation to be continuously challenged.*

DOI: 10.1057/9781137312402

So, are the technophiles convincing enough about the merits of the Digital Age and its effects on the postcapitalist marketplace? The digital technophobes on the other end of the spectrum remain unconvinced. For them, all the evils of super-capitalism – concentration of wealth and power, exploitation, alienation, and violation of privacy – are still in place, now magnified by enhanced technologies. The evils of state communism – from surveillance to planning – are likewise amplified with digital technologies whose global reach knows no bounds. Among them we find those concerned with specific networks, such as Google (Auletta 2009), and those whose concern is more widely cast in regards to social networks and the Internet (Andrews 2012). We find those alarmed by the Internet transforming us into e-personalities (Aboujaoude 2011), reducing our mental capabilities (Carr 2010), or wreaking havoc with our culture (Lanier 2010). There are those who focus on particular markets where digital technologies have caused financial disasters, as in the stock markets (Arnuk and Saluzzi 2012), while others have broader concerns about the loss of democratic ideals (Berry 2008). The list goes on with every publication of yet another sensational critique, personalizing on one level and generalizing on the other.

It seems reasonable to propose a middle-ground between the two extremes of the spectrum of reactions to the torrent of digital technologies and their gadgets. Will such a proposal seem like a copout? Will such a proposal evade the strong and valid claims made by the two extremes? Superficially the answer is yes, as if a choice – you are with us or against us – is being made. *But the very mind-set of either/or has been challenged throughout this book: postcapitalism and postmodernism exemplify attitudes that encourage both/and. In so doing, postcapitalism and postmodernism acknowledge the advantages offered by both sets of frameworks, idealized capitalism and socialism, and parlay them into multiple marketplaces that experiment and learn from their errors, improving their activities over time, whether they are related to social networks or financial markets. The open-mindedness they require demands more, not less, critical engagement with legal and ethical questions that might come up later, after a digital technology is introduced.*

Just like the card-counting gambler who is asked to leave the casino, any digital expert needs to realize under what rules the game is being played. This is true whether the intent is winning or surviving. Once the game is not limited to a casino, and once the gambling is not limited to money, as is the case with a global ecology, the stakes are much higher. What seems revolutionary right now – a real paradigm change of sorts –

DOI: 10.1057/9781137312402

may turn out to be piecemeal engineering – an ongoing process that transforms and reforms the paradigm along the way without breaking it apart (Popper 1957). If this is the case, we can reduce the hyperbole and the volume, and strive for utopian visions that are reasonable and achievable, those that require smart work (rather than hard work), patience (rather than fast results), and cooperative competition (rather than vicious winner-take-all).

DOI: 10.1057/9781137312402

Selected Bibliography

Elias Aboujaoude (2011), *Virtually You: The Dangerous Powers of the E-Personality*. New York: W.W. Norton.

Lori Andrews (2012), *I Know Who You Are and I Saw What You Did: Social Networks and the Death of Privacy*. New York: Free Press.

Julia Angwin (2009), *Stealing MySpace: The Battle to Control the Most Popular Website in America*. New York: Random House.

Arjun Appadurai (1996), *Modernity at Large: Cultural Dimensions of Globalization*. Minneapolis: University of Minnesota Press.

Sal Arnuk and Joseph Saluzzi (2012), *Broken Markets: How High Frequency Trading and Predatory Practices on Wall Street Are Destroying Investor Confidence and Your Portfolio*. New York: FT Press.

Ken Auletta (2009), *Googled: The End of the World as We Know It*. New York: Penguin.

Scott Barrett (2007), *Why Cooperate? The Incentive to Supply Global Public Goods*. Cambridge and New York: Oxford University Press.

Jean Baudrillard (1983), *Simulations*. Translated by Paul Foss, Paul Patton, and Philip Beitchman. New York: Semiotext(e), Inc.

Zygmunt Bauman (1989), *Modernity and the Holocaust*. Ithaca, NY: Cornell University Press.

Frederic L. Bender (2003), *The Culture of Extinction: Toward a Philosophy of Deep Ecology*. New York: Humanity Books.

 DOI: 10.1057/9781137312402

Walter Benjamin (1936), "The Work of Art in the Age of Mechanical Reproduction," reprinted by Prism Key Press, 2010.
Yochai Benkler (2006), *The Wealth of Networks: How Social Production Transforms Markets and Freedom*. New Haven, CT and London: Yale University Press.
David M. Berry (2008), *Copy, Rip, Burn: The Politics of Copyleft and Open Source*. London: Pluto Press.
David M. Berry (2011), *The Philosophy of Software: Code and Meditation in the Digital Age*. Hampshire, UK: Palgrave Macmillan.
Ann Blair (2010), *Too Much to Know: Managing Scholarly Information Before the Modern Age*. New Haven, CT: Yale University Press.
David Bollier (2005), *Brand Name Bullies: The Quest to Own and Control Culture*. Hoboken, NJ: J. Wiley & Sons.
James Boyle (1996), *Shamans, Software, and Spleens: Law and the Construction of the Information Society*. Cambridge, MA and London: Harvard University Press.
James Boyle (2008), *The Public Domain: Enclosing the Commons of the Mind*. New Haven, CT and London: Yale University Press.
Stephen Breyer (1982), *Regulation and Its Reform*. Cambridge, MA: Harvard University Press.
John Brockman (ed.) (2011), *Is the Internet Changing the Way You Think? The Net's Impact on Our Minds and Future*. New York: Harper Perennial.
John Seely Brown and Paul Duguid (2000), *The Social Life of Information*. Boston: Harvard Business School Press.
Jessica Bruder (2012), "What if Web Users Could Sell Their Own Data?," *The New York Times*, October 2.
Gordon Calleja (2007), "Digital Games as Designed Experience: Reframing the Concept of Immersion," PhD dissertation, Victoria University of Wellington.
Nicholas G. Carr (2010), *The Shallows: What the Internet Is Doing to Our Brains*. New York: W.W. Norton.
Ha-Joon Chang (2010), *Things They Don't Tell You About Capitalism*. New York: Bloomsbury Press.
Michael Chorost (2011), *World Wide Mind: The Coming Integration of Humanity, Machines, and the Internet*. New York: Free Press.
Andy Clark (2008), *Supersizing the Mind: Embodiment, Action, and Cognitive Extension*. New York: Oxford University Press.

DOI: 10.1057/9781137312402

Arthur C. Clarke (1985), *Profiles of the Future* [1962]. New York: Warner Books.

Antoine-Nicolas de Condorcet (1979), *Search for the Historical Picture of the Progress of the Human Mind* [1795]. Translated by June Barraclough. Westport, CT: Greenwood Press, Publishers.

Robert Darnton (2009), "Google and the New Digital Future." *The New York Review*, December 17, pp. 82–84.

William H. Davidow (2011), *Overconnected: The Promise and Threat of the Internet.* Harrison, NY: Delphinium Books.

Chris Dede (2005), "Planning for Neomillennial Learning Styles," *EDUCAUSE Quarterly* 28:1, 7–12.

Jacques Derrida (1976), *Of Grammatology* [1967]. Translated by Gayatri Chakravorty Spivak. Baltimore, MD: Johns Hopkins University Press.

Peter H. Diamandis and Steven Kotler (2012), *Abundance: The Future Is Better Than You Think.* New York: Free Press.

Peter F. Drucker (1993), *Post-Capitalist Society.* New York: HarperBusiness.

Milad Doueihi (2011), *Digital Cultures.* Cambridge, MA and London: Harvard University Press.

Charles Duhigg and Steve Lohr (2012), "The Patent, Used as a Sword," *The New York Times*, October 8.

George Dyson (2012), *Turing's Cathedral: The Origins of the Digital Universe*. New York: Vintage.

Jacques Ellul (1964), *The Technological Society*[1954]. New York: Vintage.

Charles Ess (2009), *Digital Media Ethics*. Cambridge, UK: Polity.

Milton Friedman (1982), *Capitalism and Freedom* [1962]. Chicago and London: University of Chicago Press.

Steve Fuller (2003), *Behind the Blip: Essays on the Culture of Software.* London: Autonomedia.

Adam Gopnik (2011), "The Information: How the Internet Gets Inside Us," *The New Yorker*. February 14 and 21, pp. 124–130.

Oliver Grau (2003), *Virtual Art: From Illusion to Immersion*. Cambridge: MIT Press.

Reuben Grinberg (2012a), "Bitcoin: An Innovative Alternative Digital Currency," *Hastings Science & Technology Law Journal*, 4:1 Winter, 159–208.

Reuben Grinberg (2012b), "Today Techies, Tomorrow the World? Bitcoin," *The Milken Institute Review*, First Quarter, 23–31.

DOI: 10.1057/9781137312402

Ihab Hassan (1987), "Toward a Concept of Postmodernism," *The Postmodern Turn*. Edited by Joseph Natoli and Linda Hutcheon Columbus, OH: Ohio State University Press, pp. 273–286.

Simon Head (2007), "They're Micromanaging Your Every Move," *New York Review of Books*, August 16, pp. 42–44.

Martin Heidegger (1954), "The Question Concerning Technology," reprinted in *Basic Writings*. New York: Harper & Row, 1977, pp. 287–317.

Richard Hofstadter (1955), *Social Darwinism in American Thought* [1944]. Boston: Beacon Press.

Richard Hofstadter (1966), *Anti-Intellectualism in American Life*. New York: Vintage.

Friedrich A. Hayek (1972), *The Road to Serfdom* [1944]. Chicago: University of Chicago Press.

Steven Johnson (2012), "The Internet? We Built That," *New York Times*, September 23, Magazine.

Hans Jonas (1984), *The Imperative of Responsibility: In Search of an Ethics for the Technological Age*. Chicago and London: University of Chicago Press.

Jesper Juul (2012), *A Casual Revolution: Reinventing Video Games and Their Players*. Cambridge: MIT Press.

Feroz Khan (2002), *Information Society in Global Age*. New Delhi, India: A. P. H. Publishing Corporation.

Immanuel Kant (1951), *Critique of Judgment* [1790]. Translated by J. H. Bernard. New York: Hafner Press.

Immanuel Kant (1981), *Grounding for the Metaphysics of Morals* [1785], translated by James W. Ellington. Indianapolis: Hackett Publishing Co.

Friedrich A. Kittler (1999), *Gramophone, Film, Typewriter* [1986]. Translated by Geoffrey Winthrop-Young and Michael Wutz. Stanford, CA: Stanford University Press.

Friedrich Kittler (2010), *Optical Media: Berlin Lectures 1999* [1999]. Translated by Anthony Enns. Cambridge, UK: Polity Press.

Eric Klineberg (2012), *Going Solo: The Extraordinary Rise and Surprising Appeal of Living Alone*. New York: Penguin.

Eric Kluitenberg (2008), *Delusive Spaces: Essays on Culture, Media, and Technology*. Institute of Network Cultures: NAi Publishers.

Jaron Lanier (2010), *You Are Not a Gadget: A Manifesto*. New York: Alfred A. Knopf.

DOI: 10.1057/9781137312402

Lawrence Lessig (2001), *The Future of Ideas: The Fate of the Commons in a Connected World*. New York: Vintage.

Lawrence Lessig (2004), *Free Culture: How Big Media Uses Technology and the Law to Lock Down Culture and Control Creativity*. New York: Penguin.

Lawrence Lessig (2008), *Remix: Making Art and Commerce Thrive in the Hybrid Economy*. New York: Penguin.

Brink Lindsey (2007), *The Age of Abundance: How Prosperity Transformed America's Politics and Culture*. New York: Collins.

Geert Lovink (2007), *Zero Comments: Blogging and Critical Internet Culture*. New York: Routledge.

Geert Lovink (2012), *Networks without a Cause: A Critique of Social Media*. Cambridge, UK: Polity.

Jean-François Lyotard (1984), *The Postmodern Condition: A Report on Knowledge* [1979]. Translated by Geoff Bennington and Brian Massumi. Minneapolis: University of Minnesota Press.

Jerry Mander (1978), *Four Arguments for the Elimination of Television*. New York: Harper Perennial.

Herbert Marcuse (1964), *One-Dimensional Man: Studies in the Ideology of Advanced Industrial Society*. Boston: Beacon Press.

Karl Marx (1976), *Capital: A Critique of Political Economy; Volume One* [1867] . Translated by Ben Fowkes. New York: Vintage.

Karl Marx and Frederick Engels (1970), *The German Ideology* [1932]. Edited by C. J. Arthur. New York: International Publishers.

Ernst Mayr (1997), *This Is Biology: The Science of the Living World*. Cambridge, MA: Harvard University Press.

Jane McGonigal (2011), *Reality Is Broken: Why Games Make Us Better and How They Can Change the World*. New York: Penguin.

Kembrew McLeod (2005), *Freedom of Expression®: Overzealous Copyright Bozos and Other Enemies of Creativity*. New York: Doubleday.

Kembrew McLeod and Peter DiCola (2011), *Creative License: The Law and Culture of Digital Sampling*. Duke University Press.

Marshall McLuhan (1994), *Understanding Media: The Extensions of Man* [1964]. Cambridge and London: MIT Press.

Ian F. McNeely and Lisa Wolverton (2008), *Reinventing Knowledge: From Alexandria to the Internet*. New York: W.W. Norton.

Robert K. Merton (1973), *The Sociology of Science: Theoretical and Empirical Investigations*. Chicago and London: University of Chicago Press.

DOI: 10.1057/9781137312402

Ben Mezrich (2010), *The Accidental Billionaires: The Founding of Facebook: A Tale of Sex, Money, Genius, and Betrayal*. New York: Anchor Books.

Evgeny Morozov (2011), *The Net Delusion: The Dark Side of Internet Freedom*. New York: Public Affairs.

Nakamoto, Satoshi (24 May 2009), "Bitcoin: A Peer-to-Peer Electronic Cash System," pp. 1–9. Retrieved 9/26/2012.

Martin A. Nowak (2011), *Super Cooperators: Altruism, Evolution, and Why We Need Each Other to Succeed*. New York: Free Press.

Eli Pariser (2011), *The Filter Bubble: What the Internet Is Hiding from You*. New York: Penguin.

Karl R. Popper (1957), *The Poverty of Historicism*. New York: Harper & Row.

William Powers (2010), *Hamlet's BlackBerry: A Practical Philosophy for Building a Good Life in the Digital Age*. New York: Harper.

I. I. Rabi (1970), *Science: The Center of Culture*. New York and Cleveland: The World Publishing company.

Kal Raustiala and Christopher Sprigman (2012), *The Knockoff Economy: How Imitation Sparks Innovation*. Oxford and New York: Oxford University Press.

Robert B. Reich (2007), *Supercapitalism: The Transformation of Business, Democracy, and Everyday Life*. New York: Alfred A. Knopf.

Alexander Rosenberg (1988), *Philosophy of the Social Sciences*. Boulder: Westview Press.

Tina Rosenberg (2011), *Join the Club: How Peer Pressure Can Transform the World*. New York: W.W. Norton.

Paul Rosenstein-Rodan (1961), "Notes on the Theory of the Big Push," *Economic Development for Latin America*. Edited by H. S. Ellis and H. C. Wallich. New York: St. Martin's Press.

Michael Rothfeld (2012), "Corporate Probation: Punishing or Punting?," *The Wall Street Journal*, August 31. http://online.wsj.com/article/SB100 00872396390444772804577621780469137056.html

Jean-Jacques Rousseau (1964), "Discourse on the Sciences and Arts," *The First and Second Discourses* [1750]. Edited by Roger D. Masters. Translated by Roger and Judith R. Masters. New York: St. Martin's Press, pp. 31–74.

Jean-Jacques Rousseau (1978), *On the Social Contract* [1762]. Edited by Roger D. Masters. Translated by Judith R. Masters. New York: St. Martin's Press.

DOI: 10.1057/9781137312402

Bertrand Russell (1949), *Authority and the Individual.* Boston: Beacon Press.

Richard M. Ryan, C. Scott Rigby, Andrew Przybylski (2006), "The Motivational Pull of Video Games: A Self-Determination Theory Approach," *Springer Science+Business Media*, LLC, pp. 1–17.

Jeffrey D. Sachs (2008), *Common Wealth: Economics for a Crowded Planet.* New York: Penguin.

Raphael Sassower (1993), *Knowledge Without Expertise: On the Status of Scientists*. Albany, NY: SUNY Press.

Raphael Sassower (1995), *Cultural Collisions: Postmodern Technoscience.* New York and London: Routledge.

Raphael Sassower (1997), *Technoscientific Angst: Ethics and Responsibility.* Minneapolis and London: University of Minnesota Press.

Raphael Sassower (2009), *Postcapitalism: Moving beyond Ideology in America's Economic Crises*. Boulder and London: Paradigm Publishers.

Raphael Sassower (2012), "We Need to Raise Ethical Expectations," *The Colorado Springs Business Journal*, September 14–20, 21.

Richard Sennett (2006), *The Culture of the New Capitalism*. New Haven, CT: Yale University Press.

Clay Shirky (2010), *Cognitive Surplus: Creativity and Generosity in a Connected Age*. New York: Penguin.

Georg Simmel (1978), *The Philosophy of Money* [1900]. Translated by Tom Bottomore and David Frisby. London: Routledge & Kegan Paul.

Natasha Singer (2012), "Secret E-Scores Chart Consumers' Buying Power," *The New York Times*, August 18.

Adam Smith (1937), *An Inquiry into the Nature and Causes of the Wealth of Nations* [1776]. Edited by E. Cannan. New York: The Modern Library.

Adam Smith (1976), *The Theory of Moral Sentiments* [1759]. Indianapolis: Liberty Classics.

C. P. Snow (1959), *The Two Cultures and a Second Look.* Cambridge: Cambridge University Press.

Betsy Sparrow, Jenny Liu, Daniel M. Wegner (2011), "Google Effects on Memory: Cognitive Consequences of Having Information at Our Fingertips," *Science* 333, pp. 776–778.

Joseph E. Stiglitz (2003), *Globalization and Its Discontents.* New York and London: W.W. Norton.

Don Tapscott and Anthony D. Williams (2010), *MacroWikinomics: Rebooting Business and the World.* London: Atlantic Books.

DOI: 10.1057/9781137312402

Eric Topol (2012), *The Creative Destruction of Medicine: How the Digital Revolution Will Create Better Health Care*. New York: Basic Books.

Michel-Rolph Trouillot (1995), *Silencing the Past: Power and the Production of History*. Boston: Beacon Press.

Edward R. Tufte (1983), *The Visual Display of Quantitative Information*. Cheshire, CT: Graphics Press.

Sherry Turkle (2011), *Alone Together: Why We Expect More from Technology and Less from Each Other*. New York: Basic Books.

Ludwig von Mises (1984), *The Historical Setting of the Austrian School of Economics* [1969]. Auburn, AL: The Ludwig von Mises Institute of Auburn University.

Edward O. Wilson (2012), *The Social Conquest of Earth*. New York and London: W.W. Norton.

Langdon Winner (1977), *Autonomous Technology: Technics-Out-of-Control as a Theme in Political Thought*. Cambridge and London: MIT Press.

www.theesa.com/facts; cited 6/25/2011

DOI: 10.1057/9781137312402

Index

Aboujaoude, Elias/e-personality, x, 14–15, 70
Arab Spring, 29–30, 36
Austrian School of Economics, 36–9, 45

Barrett, Scott/global cooperation, 34, 69
Bauman, Zygmunt/modernist rationality, 13–14
Benkler, Yochai/network economy, 26–8, 30, 33, 52, 54
Berry, David/political critique of digital technologies, 2, 8, 52, 53–5, 58, 70
  computational rationality, 8
  digital environmentalism, 54
Boyle, James/legal critique of digital technologies, vii–viii, x, 2, 13, 54

Carr, Nicholas/Internet influence, 15–16, 70
commons, 24, 29–31, 52, 54–5, 56–7
credit score/e-score, 43–4

Davidow, William/interconnectedness, 24–5, 32, 45
Diamandis, Peter/abundance, 67–8
Doueihi, Milad/digital divide, 12–13
Drucker, Peter/post-capitalism, 21

economic collusion, 28–9
Ellul, Jacques/Technological age, 12–13
Enlightenment, vii–ix, 9, 11, 16–17, 28, 64

Frankfurt School, 9, 10, 12, 13

globalization, 31–4, 54–5, 67–70
Gopnik, Adam/responses to digital technologies, 17

Hassan, Ihab/postmodernism, 2–3
Hayek, Friedrich/Austrian School, 38–9
Head, Simon/enterprise systems, 40–1
Heidegger, Martin/technology, 8, 9, 15
high-frequency trading, 47, 61, 63
Hofstadter, Richard/anti-intellectualism, social Darwinism, 6, 22

intellectual property, x, 2, 25, 30, 52–6, 61–2, 69

Johnson, Steven/mega-projects, 27–8

DOI: 10.1057/9781137312402

Kant, Immanuel/ethics, 11, 61
Kittler, Friedrich/technology, warfare, 6–8, 9–10, 11
Kotler, Steven/abundance, 67–8

Lanier, Jaron/cybernetic totalitarianism, ix, 14, 70
Lawrence, Lessig/limits to copyright, x, 52–3, 54
Lyotard, Jean-Francois/postmodernism, x, 2, 11, 56

Marx, Karl/Marxism, vii–ix, xi, 20–1, 26, 32–4, 38, 68
McLuhan, Marshall/media, 12–13
Merton, Robert/scientific ethos, 68
Morozov, Evgeny/cyber-utopianism, Internet-centrism, ix, 16

Nowak, Martin/models of cooperation, 23–4

Pariser, Eli/consumer data collection, 16, 41–2
Popper, Karl/Popperianism, vii, 26, 70–1
postcapitalism, viii, xi, 4, 17, 20–1, 23, 32, 50, 53, 61–3, 70–1
postmodernism, vii, x–xi, 2–4, 9–11, 14, 17, 50, 56–7, 59–60, 69–71

Rabi, I. I./wisdom, 5, 7
Raustiala, Kal/knockoff economy, 69
reductionism, x, 6, 8, 14
Reich, Robert/super-capitalism, 54
Rosenstein-Rodan, Paul/big-push development, 33–4
Rousseau, Jean-Jacques/social contract, 39, 64–5
Russell, Bertrand/role of government, 64

Sachs, Jeffrey/globalism, 32–4
Schumpeter, Joseph/creative destruction, xi–xii
Shirky, Clay/cognitive surplus, 16, 48
Simmel, Georg/money, 57–8
Singer, Natasha/ Internet commerce, 43–5
Smith, Adam, viii, x–xi, 20, 24, 25, 26, 31–2, 33, 37
Snow, C. P./two cultures, 5, 7, 9, 13
Sprigman, Christopher/knockoff economy, 69
Stiglitz, Joseph/globalism, 31–2
surveillance, x, 29, 36, 39–46, 59, 70

Tapscott, Don/collaborative communities, 68
technophiles/technophobes, ix, 8–18, 67–70
Topol, Eric/digital medicine, xi–xii, 55

venture capital/angel investors/crowd-funding, 46–50
virtual reality/space, 7, 14–15, 25, 30, 53, 57, 59
virtual currency, 56–60
Von Mises, Ludwig/Austrian School, 37–9

Williams, Anthony/collaborative communities, 68
Wilson, Edward O./multilevel evolution, 22–4, 34, 37, 69

DOI: 10.1057/9781137312402

CPSIA information can be obtained at www.ICGtesting.com
Printed in the USA
LVOW11*0902221214

419931LV00008B/64/P

9 781137 312396